Future Sea

FUTURE SEA

HOW TO RESCUE AND PROTECT
THE WORLD'S OCEANS

Deborah Rowan Wright

The University of Chicago Press
Chicago and London

The University of Chicago Press, Chicago 60637
The University of Chicago Press, Ltd., London
© 2020 by Deborah Rowan Wright
All rights reserved. No part of this book may be used
or reproduced in any manner whatsoever without
written permission, except in the case of brief
quotations in critical articles and reviews. For more
information, contact the University of Chicago Press,
1427 E. 60th St., Chicago, IL 60637.
Published 2020
Printed in the United States of America

29 28 27 26 25 24 23 22 21   2 3 4 5

ISBN-13: 978-0-226-54267-6 (cloth)
ISBN-13: 978-0-226-54270-6 (e-book)
DOI: https://doi.org/10.7208
/chicago/9780226542706.001.0001

Library of Congress Cataloging-in-Publication Data

Names: Wright, Deborah Rowan, author.
Title: Future sea : how to rescue and protect the
world's oceans / Deborah Rowan Wright.
Description: Chicago : The University of Chicago
Press, 2020. | Includes bibliographical references and
index.
Identifiers: LCCN 2020024360 | ISBN 9780226542676
(cloth) | ISBN 9780226542706 (ebook)
Subjects: LCSH: Marine ecosystem health. | Marine
habitat conservation—Government policy. | Marine
habitat conservation—Law and legislation.
Classification: LCC QH541.5.S3 W75 2020 | DDC
577.7—dc23
LC record available at https://lccn.loc.gov/2020024360

♾ This paper meets the requirements of
ANSI/NISO Z39.48-1992 (Permanence of Paper).

FOR ELLIE, LYDIA, AND BETTY

Contents

Introduction

Steep green slopes of red cedar and Douglas fir tumble down to Johnstone Strait, a stretch of the Pacific separating Vancouver Island from mainland Canada. This place is a gift for the eyes and the heart: an unruly expanse of forest cradling a flickering sea, with a sweet, clean chill in the air. The water is deep, and I am perched precariously upon it in a small boat, setting out on a whale-watching trip with my two daughters and my niece. We are heading south from Telegraph Cove, leaving the pale peaks of the Pacific Ranges melting into the distance beyond Blackfish Sound.

I stand swaying on the deck, feeling inconsequential in the hugeness of this land, this sea, and this sky. Wayward strands of wet hair cling to my cold face. Against the headwind we look over the bow and down into the shifting darkness below, wondering. Today we hope for a lot. They say there's no better place in the world to see the magnificent orcas.

Our fellow passengers stay in the cabin, sheltering from the drizzle and the wind. Up in the wheelhouse, Captain Wayne is at the helm and biologist Kyle is at the microphone, talking about waterproof trousers, hot coffee, and salmon runs. His voice is snatched away by the bluster outside and I catch only bits, so I listen harder. He's listing the wildlife we're likely to see. I hear "bald eagles, humpback whales, dolphins, sea lions" and, last, "orcas—otherwise called killer whales." Kyle explains that orcas aren't whales at all, but the largest species of dolphin. They live in complex matriarchal family groups, with up to four generations in a single pod. There are two distinct types in this part of British Columbia: the transient Bigg's orcas that roam widely along the Pacific coast, feeding on marine mammals such as dolphins and seals, and the resident orcas with an entirely different diet, primarily Chinook salmon. My mind drifts to what could be swimming under us. Will the fabled strait deliver?

A flurry of Bonaparte's gulls gather above, on the lookout for lunch, as we make for the Michael Bigg Ecological Reserve along the coast at Robson Bight. This is a small chunk of protected land and sea that is off-limits to the public. It includes a swath of water and foreshore where orcas have come for generations to socialize and rub their bodies on the pebbles to scrape off barnacles. We're told there's a good chance of seeing them here. Suddenly the three girls squeal, bouncing and pointing like maniacs. They've spotted a surge of Pacific white-sided dolphins ahead, weaving through the water in unison, arching and diving in a lather of speed. We jump up, beckoning the others to leave the cabin, but within a few moments the dolphins are gone, leaving behind only trails of white water.

As our boat nears the reserve Captain Wayne cuts the engine so as not to disturb the animals, and we rock to the rhythm of the waves. Everyone is out on deck now, all eyes fixed on a stretch of water where the ridge falls into the bight, many with binoculars pressed to their faces. We wait and wait, watching expectantly. The rain lets up and clouds dissolve to reveal a backcloth of blue sky, but the orcas don't come. Rumbling disappointment is quickly dispelled, though. Kyle announces that the captain of another whale-watching boat has alerted him to a pod of orcas foraging for fish in the riptide about a mile farther south. We set off immediately.

Within a few minutes we see a strip of troubled water ahead, effervescent with currents clashing below. A great number of noisy birds dive to pick off fish caught in the fighting tides: rhinoceros auklets, common guillemots, and a variety of gulls. The boat's engine goes quiet and we stop, bobbing on the swell. The first sign that orcas aren't far away is a tall column of water vapor shooting up near the shore. Soon after, much closer to us, a glossy dorsal fin pierces the surface to yelps of joy on deck. Then, without warning, another orca emerges right next to the boat with a roaring "swoosh" as he breathes out, sending up a salvo of spray that makes us all gasp. It's a large male with a dorsal fin five or six feet tall. He glides past effortlessly on the port side and slips back down—six tons of bone, brains, and blubber wrapped in a mighty black submarine body.

By the time people move across the deck to watch, more orcas appear on the other side—a group of four, including a mother and calf. They are leaving the riptide banquet in a playful mood, circling each other, scribbling patterns in the water. They slap their tails down hard, making a distinct "crack" as the flukes hit the sea. The big male joins them and rolls on his side, lifting a pectoral fin as if to greet his family with a high five. We cheer and clap. Three more orcas rise together, their fins slicing the waves like knives cutting through rippling silk. People hardly know where to stand, afraid of missing something. The dolphins are back. They move toward us at a frantic pace, some leaping clear of the water, churning the surface into a froth as they swerve around the orcas. But now the orcas begin to head off, and the spectacle is coming to an end—not before an encore, though. The last to go is a large female. Suddenly she rises upright to get a good look at us, with a third of her body clear of the water. After a few moments her curiosity is satisfied. Obviously we aren't all that interesting. She sinks back down and is gone, leaving us jumping, laughing, and clapping like excited children.

A posse of boisterous Dall's porpoises escorts us back to Telegraph Cove. They are surfing in the wake of the bow with frenzied energy, so close to the keel that when I lean over the side I can almost touch them. A bald eagle breezes overhead as we approach the harbor. It's the end of the trip. I gaze wistfully along the strait, then smile as I decide to do it all again the next day.

This region is rich in marine life. The inlets and coves, channels and deep fjords provide unique habitats and diverse ecosystems for a wide range of species: plants, fish and crustaceans, birds and mammals. A combination of ocean currents, eddies, and upwellings makes the waters very productive, and so much food brings in a host of hungry predators. But despite the high probability of seeing orcas in Johnstone Strait, the northern resident population is quite small—roughly three hundred individuals—and there are only about seventy southern resident orcas, which live in the Salish Sea about two hundred miles farther south. Scientists are unclear on why the number of orcas remains low, and various explanations have been put forward. During the 1960s and 1970s many orcas were captured and taken to zoos and marine parks like SeaWorld. They are long-lived animals and slow to reproduce, and it takes many years for their numbers to recover. Thankfully Vancouver Island's orcas are protected by Canada's Species at Risk legislation and can no longer be whisked away into captivity, but there are other threats to their long-term well-being. Waterborne and airborne pollutants from agriculture and industry known as persistent organic pollutants reach the sea and contaminate the fish that orcas and other marine mammals eat, undermining the animals' health and strength. The concentration of these toxins in the body increases at every stage of the food chain, so top predators like orcas have the highest levels.

Underwater noise from ships and submarines and the use of sonar interfere with the orcas' ability to use echolocation to navigate, hunt, and communicate with one another. Whales, dolphins, and orcas become distressed, show signs of pain, and may become stranded on the shore. This happens frequently when naval exercises are underway in the vicinity. The greatest problem they face, however, is simply a shortage of food. Although orcas may be protected, sadly their food is not. Chinook salmon stocks have been shrinking for years, and again the reason is open to question. Marine biologists argue that large-scale commercial fishing takes too many of the fish the orcas and so many other species depend on. Poor logging management and dam building along the region's rivers have destroyed critical salmon spawning areas, significantly reducing their reproduction rates. In 2019 the government department Fisheries and Oceans Canada introduced tougher restrictions on chinook fishing in British Columbian waters, closing some recreational and commercial fishing areas in the region to try to get the fish back. Southern resident orcas in the Salish Sea have been particularly affected. By 2018, no successful births had been recorded during the previous three years, so it was a relief to many when in the summer of 2019 two newborn calves were spotted swimming alongside their mothers off the coast of Vancouver Island.

Seldom is it possible to be certain why a wild species is in decline, and such declines are generally due to a complex mix of several factors. What we know about life under the waves is outweighed by what we don't know—the certainty of uncertainty. Recognizing this should be at the heart of our relationship with the ocean and the way we use the sea's resources. Such an approach provides a safety net for misjudgment when science and policy have insufficient knowledge or when they are mistaken. This safety measure thinking—known as the precautionary approach—is an essential element of the possible future for the oceans that I describe in the chapters to come.

All my expectations of seeing orcas in Johnstone Strait were easily surpassed. I'll never forget those two sublime days of adventure on the water. Being so close to animals in the wild is an exhilarating experience, and for most of us it is an uncommon one. We have become detached from the wild places we inhabited long ago and the from the wild creatures living around us. We typically find ourselves enclosed in static environments of concrete and glass laced by roads, not rivers, and dodging cars and trucks rather than fleeing wolves or tigers. For some, a connection with nature is limited to a walk in the park and watching the occasional wildlife documentary on television. Often we suffer from an overload of bad news about what's happening to the natural world, and we switch off or turn the page when the latest crisis is reported. Thinking there is little or nothing we can do, we hope someone else will deal with it or decide it's not our problem anyway. In fact, there are constructive things that anyone can do to offset the gloom, and there are plenty of reasons to feel optimistic. This book brings good news and lays out solutions that everyone can be a part of. It is a proposal for a radical course of action to bring back fish-filled seas, to clean polluted waters, protect underwater habitats, and safeguard the oceans' astonishing wildlife for good.

1: Back-to-Front World

In the deepest, darkest waters of the world's oceans, life moves along at a leisurely pace. Here many fish species grow slowly and mature late. They produce relatively few offspring and live to a great age if they're not eaten by predators or caught in a deep-sea trawl net. Take the slimehead, which doesn't reach reproductive age until it is 30 to 40 years old and can live 150 years or possibly more. Using radiometric dating of trace isotopes from the ear bone, one individual was found to be 149 years old.[1] The fishing industry had overlooked slimeheads as having no commercial value, which meant they were still plentiful around many undersea mountains and valleys. But as shallow-water fish stocks have become exhausted, vessels head farther out to sea, fishing in deeper waters and targeting species that previously were safe from overfishing. In 1979 Russian trawlers chanced upon large populations of slimeheads in deep seas to the east of New Zealand.

Unfortunately for slimeheads, their flesh is firm and white, with a mild and unchallenging flavor that most people apparently prefer. So with these two winning characteristics—blandness and abundance— they were deemed ideal for the consumer and ripe for exploitation. With rapid and easy profits to be made, fisheries marketers aimed to get them onto restaurant tables and into supermarkets around the world. Their promotion plans met a minor stumbling block, though. "Whole-roasted slimehead with lemon dressing" would hardly be a restaurant's most popular dish. The answer was a simple rebranding, and the fish were given the more appealing name orange roughy, after their fiery color. And so began the scramble for roughy—an under- water gold rush that lasted throughout the 1980s and into the 1990s, mainly in seas around Australia and New Zealand. The bonanza peaked in 1990 when the global catch of orange roughy topped 91,000 tons.[2] Their combination of a long life and late maturity makes this species acutely vulnerable to excessive fishing. Roughy could not reproduce fast enough to keep pace with the huge numbers being taken, and before long catches nose-dived. Within twenty years some populations were estimated to be only 3 to 10 percent of their previous size. Once breeding fish are gone the stock cannot rebuild for decades, if at all. Fisheries managers had used guesswork instead of facts to set catch quotas. They took what they knew about similar-sized species living in shallow seas and applied it to the orange roughy, whose life cycle is up to ten times longer and slower. The damage was made worse by the use of weighted nets that destroy slow-growing corals and sponges found in areas where the fish gather to feed and spawn.

The orange roughy's story is not just another case of greedy boom-and-bust fishing. It shows how devastating it can be when management decisions are based on patchy knowledge or mistaken assumptions. It reveals why understanding, caution, and moderation should underpin everything we do in the sea. If fisheries management of orange roughy had been led from the outset by the precautionary approach, there would still be great clouds of them swimming over undamaged ancient corals in the deep seas of the southwest Pacific. Because we choose not to take the precautionary route, other species are suffering a similar demise today.

California sea lions live along the Pacific coast from Vancouver to Mexico. They are inquisitive, playful, and sociable animals, and people love to watch them racing and twisting through the water or just relaxing in groups on the rocks and the foreshore. But in recent years large numbers of pups have been washed up on West Coast beaches, starving and dying. Sardines and anchovies are a key part of the sea lions' diet, and as stocks have declined mothers have to go farther and farther in search of food for their young. Figures from America's National Marine Fisheries Service indicate a 98 percent collapse of the Pacific sardine population since 2006, which means that pups starve or drown when they take to the sea before they are mature enough to survive. Sardines are a crucial forage food for many other species such as dolphins, whales, brown pelicans, marlins, and tuna. Scientists, conservationists, and those in the fishing industry disagree about why stocks collapsed. Some say warming seas are sending the fish north to cooler waters, while others put it down to poor fisheries management that has set catch quotas too high for too long and allowed commercial fishing to continue beyond biologically safe limits.

The plight of British Columbia's orcas, of orange roughy, and of California's sea lions illustrates the difficulties facing wildlife in many places: seas emptied of fish, underwater habitats flattened, waters polluted and progressively altered by climate change. But I don't intend to dwell on the many threats to the oceans—I intend to set out a way to eliminate them. Fortunately there are hundreds of counteractive strategies and initiatives already protecting marine environments, and many are very successful. Simply managing fisheries wisely makes a massive difference. This might be done on a national scale, exemplified by countries such as Norway and New Zealand, or it can occur on a smaller scale, with community-led conservation projects safeguarding marine resources in coastal waters. However, the strategy widely regarded as most successful is to create marine protected areas (MPAs) and marine reserves, like the Michael Bigg reserve in Johnstone Strait where we went whale watching. In these designated areas fishing, dumping, mining, dredging, and coastal development are prohibited or restricted. Marine reserves are areas with the highest level of protection, where habitats particularly rich in biodiversity are kept intact. When left undisturbed by humans, damaged habitats regenerate, and depleted populations of wildlife bounce back. As well as the ecological benefits, marine reserves bring considerable economic and social advantages for nearby communities. Older, larger fish tend to produce far more eggs than smaller, younger ones. Providing a safe haven where females can live well past maturity allows stocks to replenish. An upsurge of sea life spilling over the boundaries boosts the catches of local fisheries in adjacent waters. In time, everything and everybody wins.

There are thousands of protected areas of ocean around the world, offering some degree of sanctuary for marine life, and a number of very large MPAs are being designated, but many protected areas are too small or too detached to be very effective, or they may have only seasonal closures during spawning times to help a dwindling population rebuild. Some are "protected areas" in name only and provide no more respite than the unprotected sea. It could be that regulations are not sufficiently monitored and enforced or that there is no management plan in place. It could also be that the restrictions on human activities are so minimal that they make no real difference. And even though creating MPAs and marine reserves has proved extremely valuable, according to the American-based nongovernmental organization the Marine Conservation Institute, less than 6 percent of the global marine environment is in managed protected areas, with only about half of that fully protected and completely safe from misuse. Clearly, there are not enough well-managed marine protected areas.

The best protected areas reconcile the need to safeguard nature with people's need to use marine resources. Typically they include a "no-take" zone to be left undisturbed by humans (which might be where fish spawn or where there is a particularly fragile habitat), while a larger part of the area will be open to varying degrees of use or "harvesting"—but only when done in a responsible way that doesn't damage habitats (the reef or kelp forest, for instance), kill or injure nontarget wildlife, outstrip the target resource (such as a stock of fish), or pollute the water, the shore, or the seabed. Striking that balance between protection and exploitation means that well-managed marine protected areas provide the working model for how to keep seas loaded with life. Therefore we already know what needs to be done to allow oceans to thrive: designate more protected areas and marine reserves. But why safeguard parts of the ocean and leave the rest still vulnerable to damaging industries and unregulated commercial exploitation? The five named oceans (the Arctic, the Atlantic, the Indian, the Pacific, and the Southern) make up one continuous body of water called the global ocean (or the world ocean), and protection targets vary. Conservation organizations and some governments aim to have 10 percent, 20 percent, or 30 percent of the global ocean in protected areas. But even at best this would still leave 70 percent of the ocean open to operations like overfishing, mining in sensitive habitats, and dumping waste. Patch protection doesn't solve the fundamental problem—which is a combination of bad practice, entrenched habits, and outdated attitudes toward the natural world—it just moves the problem to a different place.

Across the world, fishers, scientists, conservationists, divers, journalists, and concerned citizens are fighting to save a critical global resource, even though healthy seas sustain all life on Earth. They work relentlessly to keep chunks of sea safe in protected areas. Meanwhile those who empty, pollute, and profit from the seas regularly do so unchallenged and uncontrolled, often with little accountability. It's a back-to-front scenario that is neither logical nor just.

Oceans make all life possible and enable it to flourish. They generate oxygen, absorb excess heat trapped by greenhouse gases, and take up millions of tons of carbon dioxide from the atmosphere. Undersea currents regulate global temperatures and climate systems, distributing the sun's heat, moderating extreme temperatures, and keeping the land habitable. New technologies can harness perpetual clean energy from the sea to replace the burning of fossil fuels. Recently discovered marine organisms are being developed to produce new antibiotics and drugs to combat cancer, HIV, and malaria. Roughly two hundred million people are directly or indirectly employed in fisheries, and the World Wildlife Fund estimates that approximately three billion people depend on marine fish as their main source of protein. But the most extraordinary feature of the world's oceans is the life within them. The variety and quantity of plant and animal species in the sea is almost beyond comprehension—estimated by the United Nations to be between 50 and 80 percent of all plants and animals on Earth.[3] Oceans give us food, livelihoods, state revenue, means of transport, energy, medicine, and recreation and are a source of beauty and inspiration. They provide habitable climates and even the air we breathe. We cannot survive without the sea in a healthy state. Knowing that, ask yourself this: Why should some in society have to justify conserving ocean life while others do not have to justify destroying it? Surely the basic premise of how we use the sea should be turned around. The onus should be on sea users to demonstrate that they do not ruin undersea habitats, contaminate waters, or jeopardize wildlife popu-lations—not on conservationists to prove why oceans and marine life need to be protected.

We already have the formula for regaining clean and healthy oceans—creating more marine reserves and protected areas—and this approach can be rolled out around the globe. Our perception of the world's oceans and the way we use their resources can be turned from default profit and exploitation to default care and respect, so that all our dealings with the sea are built on a simple universal prin-ciple: to safeguard *all* of it and use *all* of it responsibly. With all the sea protected from harm, instead of scattered oases of safe water within emptied oceans, there will be areas of responsible use within healthy oceans; Marine Commercial Areas instead of Marine Protected Areas. We could still fish the waters, ship cargo across them, and take min-erals and oil from beneath the seabed, but only in a way that safe-guards marine habitats and populations of sea life.

In the following chapters I sketch out the future we can expect if we transform our approach to the sea in favor of nature, always taking a precautionary approach to how we use marine resources—everywhere. And the good news is that the essential foundations to make this happen already exist.

2: The Laws of Life

A couple of years before the whale-watching trip in Canada, when I was camping at a music festival in the Dorset hills in the west of England, the notion to protect the whole of the ocean fell into my mind one night. It woke me up and wouldn't let me go back to sleep. Suddenly it seemed we'd become so accustomed to living in an upside-down world that we were oblivious to it. We were tolerating a system driven by an endless quest for profit that leaves much of the world's waters exposed to brutal and destructive commercial use. Oceans need to be left alone—not completely, that isn't possible—but enough to allow water and wildlife to recover and return to the way they were before we took out the good stuff and put in the bad. But was it possible to turn things right around in all of the world's seas? To turn from "default exploitation" to "default respect"? As I sank back to sleep my thoughts were drifting toward law. Working in marine conservation organizations, I'd learned something about law and the environment, and I'd picked up more from newspapers and from radio and TV reports. But my understanding was sketchy. Once I was home from the festival, I began hunting for useful laws, and what I discovered amazed me.

The next day my friend Clare came for lunch, so I sounded her out on the concept of protecting it all. "Protect the whole lot?" she said, "all of the sea, all of the ocean? Everywhere in the world? But how? People won't stop fishing."

"They don't have to stop fishing. Protection doesn't necessarily mean you can't use something. It can also mean using something well," I replied. She came back with the impossibility of it, with the overwhelming forces of business and industry, of powerful corporations and spineless governments. I said the international community would have to pass laws to keep all of the ocean safe, not just parts of it.

"Right. Like that's going to happen," she said with a cynical shrug. "They'll never do it." I only grinned.

"They will never do it," she repeated. "And why are you smiling like that?"

"They've done it," I said calmly. "Virtually all the seas and oceans in the whole world are already protected by international law."

My friend screwed her brow into an incredulous furrow and was uncharacteristically at a loss for words. In the past fifty years or so there have been over five hundred international agreements on the environment, ranging from protecting the ozone layer and Antarctica's ice to sharing genetic resources and reducing deforestation (also called multilateral environmental agreements, or MEAs). Several relate to marine conservation. It's a muddled mosaic of regulations covering various bodies of water and particular wild species, habitats, and human activities, but one is unequivocal in its stipulations to safeguard the whole marine environment and the life within it: the United Nations Convention on the Law of the Sea (UNCLOS III).

Early that morning I had found the text of the whole treaty online and scanned its extensive list of contents, which includes rules about military activities, rights of access for landlocked states, shipping, and territorial zones. Then I read section 2 of article 61, titled "Conservation of the Living Resources": "The coastal State, taking into account the best scientific evidence available to it, shall ensure through proper conservation and management measures that the maintenance of the living resources in the exclusive economic zone is not endangered by over-exploitation." (The exclusive economic zone, or EEZ, reaches two hundred miles seaward from a country's coastline, or to the edge of the continental shelf. Within the EEZ, each country has jurisdiction over the use and exploitation of the sea's resources.) Article 118 requires states to "cooperate with each other in the conservation and management of living resources in the areas of the high seas." (This is the rest of the ocean and seabed, seaward of each country's EEZ, also known as "areas beyond national jurisdiction.")

And there's more. Article 192, "States have the obligation to protect and preserve the marine environment," and article 194, section 1, "States shall take all measures . . . that are necessary to prevent, reduce and control pollution of the marine environment from any source," and section 5, "The measures taken shall include those necessary to protect and preserve rare or fragile ecosystems as well as the habitat of depleted, threatened or endangered species and other forms of marine life," and article 197, "States shall cooperate on a global basis and on a regional basis, . . . in formulating international rules, standards and recommended practices . . . for the protection and preservation of the marine environment."[1]

In theory, even though ocean acidification wasn't recognized when the treaty was drawn up, the Law of the Sea also provides a legal guard against its onset, caused by burning large quantities of fossil fuels. Article 212 of the Law of the Sea requires states to "adopt laws and regulations to prevent, reduce and control pollution of the marine environment from or through the atmosphere."

Oceans absorb huge amounts of carbon dioxide from the atmosphere, and when it dissolves in the sea carbonic acid is formed. As CO_2 emissions rise, the acidity of seawater increases (and the alkalinity is reduced). Since the Industrial Revolution began (about 250 years ago) the pH of ocean water has dropped 0.1 units, representing a 28 percent rise in acidity.[2] The full effects of ocean acidification are uncertain, but it is known to inhibit the growth of the shells and skeletons of many species, including hard corals. This could have drastic consequences for whole ecosystems, for biodiversity, and even for the planet's primary life-support systems. Coral reefs are especially vulnerable, and they have the world's greatest overall diversity of life. The International Programme on the State of the Ocean (IPSO) describes three key factors—the "deadly trio": the level of dissolved oxygen in seawater, increased acidity, and ocean warming.

The United Nations Fish Stocks Agreement (UNFSA, 1995), is an additional "sequel" treaty to the Law of the Sea that was agreed on to address inconsistencies in the original legislation on fish species that cross national boundaries or migrate long distances. The Fish Stocks Agreement was a major step forward in setting up a modern strategy for the long-term conservation of fish stocks. In the introduction the treaty calls for states to "avoid adverse impacts on the marine environment, preserve biodiversity, maintain the integrity of marine ecosystems and minimize the risk of long-term or irreversible effects of fishing operations." More specifically, in article 5(d) states are required to "assess the impacts of fishing, other human activities and environmental factors on target stocks and species belonging to the same ecosystem or associated with or dependent upon the target stocks," and in article 5(f) they are required to "minimize pollution, waste, discards, catch by lost or abandoned gear, catch of nontarget species, both fish and non-fish species, and impacts on associated or dependent species, in particular endangered species." Article 5(g, h) obliges states to "protect biodiversity in the marine environment" and to "take measures to prevent or eliminate overfishing and excess fishing capacity."

Significantly, article 6(1) of the treaty stipulates that "states shall apply the precautionary approach widely to conservation, management and exploitation of straddling fish stocks and highly migratory fish stocks in order to protect the living marine resources and preserve the marine environment" (defining the precautionary approach thus: "The absence of adequate scientific information shall not be used as a reason for postponing or failing to take conservation and management measures").[3]

If all the states fishing beyond their national jurisdictions actually complied with UNFSA's provisions, to a large extent, the high seas would be safe from harm.

The Convention on Biodiversity (CBD, 1992), aims to conserve Earth's biological diversity in a sustainable and equitable way by preserving natural habitats and systems, wild species, and genetic resources. (*Sustainable* is a term whose meaning has become unclear through overuse. The *New Oxford Dictionary of English* defines it as "conserv-

ing an ecological balance by avoiding depletion of natural resources.")
One hundred and ninety-six states have signed and ratified the treaty
(only the Vatican, Andorra, and the United States haven't done so).
In article 8(c), countries are required to "regulate or manage biological resources important for the conservation of biological diversity
whether within or outside protected areas, with a view to ensuring
their conservation and sustainable use."[4] These elements, together
with the Convention on Biodiversity's principles of the precautionary
and ecosystem approaches, put nature conservation firmly above the
demands of business and industry in the eyes of the law.

Other global codes and treaties stipulate more general ocean protection: for example, the UN World Charter for Nature (1982) and the
UN Framework Convention on Climate Change (UNFCCC, 1992).
The objective of the UNFCCC is to "stabilize greenhouse gas concentrations in the atmosphere at a level that would prevent dangerous
anthropogenic interference with the climate system." Considering
the vital role oceans play in absorbing heat and climate-changing
gases, article 4(d) of the UNFCCC is pertinent because it commits
all parties to "promote sustainable management, and promote and
cooperate in the conservation and enhancement, as appropriate, of
sinks and reservoirs of all greenhouse gases not controlled by the
Montreal Protocol, including biomass, forests and oceans as well as
other terrestrial, coastal and marine ecosystems."[5]

The 195 countries that signed the UNFCCC's Paris Agreement on
climate change in 2016 committed to step up actions and invest more
effort and resources into developing a low carbon future. The treaty
lowered the target of a maximum global temperature rise to 1.5°C
above preindustrial levels and was hailed as a great achievement
in international cooperation. Higher levels of carbon dioxide in the
atmosphere are making oceans more acidic and unable to absorb as
much of the gas as they have in the past, which makes ocean acidification a climate change issue. Therefore international efforts to reduce CO_2 emissions will be good for oceans as well as for the atmosphere. (However, implementing the Convention's agreements has
been fraught with difficulties, and governments are criticized for not

doing enough to reach their domestic targets in lowering greenhouse gas emissions, particularly carbon dioxide.)

These agreements are also overlapped by provisions for specific species, habitats, or regions, such as the those of the International Whaling Commission, the Antarctic Treaty, and the Convention on International Trade in Endangered Species of Wild Fauna and Flora (CITES).

So there it was. Almost all (about 96.5 percent) of the world's seas and the life within them are covered several times over: the waters within each country's exclusive economic zone (EEZ) plus the high seas beyond (96.5 percent comes from deducting the 3.5 percent of the countries' EEZs that have not signed the Law of the Sea). In other words, practically the entire marine world is protected by international law. That means the legal principle to protect the whole lot already exists—which is very surprising to most people.

About this time I was working on the campaign to reform the European Union's fisheries policy with Marinet, a United Kingdom–based not-for-profit organization that acts as a kind of watchdog checking that the government sticks to its marine conservation commitments and making it public when it doesn't. We were also keeping an eye on the slow progress made in establishing a network of marine conservation zones (MCZs) around British coasts, as the government was legally bound to do. At the time, the initial figure of 127 planned zones had been whittled down to a mere fifty, and dredging and fishing were still permitted in most of them. Also, the plan to create sixty-five more highly protected areas providing complete respite from human disturbance was abandoned altogether. It was another example of those in authority wriggling out of doing the right thing. Forty-one more MCZs have since been designated in UK waters, which is a positive development, but there is concern that insufficient management plans and low levels of monitoring may not deliver genuine protection.

The disappointing tale of the UK's early attempts at safeguarding its rich waters had sown seeds of skepticism in my mind. I wondered how such a pocket-size exercise in damage limitation could ever be enough to rescue the sea. Over the years before then, I had been involved in several campaigns to defend the underwater world and its beleaguered sea life with Greenpeace, Whale and Dolphin Conservation, and Marinet. We had discussed reform, reserves, and recovery. We went to windswept beaches to count migrating porpoises, attended fund-raisers dressed as mackerel, and visited food festivals to hand out flyers about unscrupulous seafood companies. We heard speakers on subjects from illegal fishing in Ghana and conservation projects in Chile to biodiversity in the Ross Sea. I wrote copy for leaflets, an academic paper for the International Whaling Commission, and scripts for short films. I compared the scientific advice on yields for European fisheries with the extravagant catch quotas fisheries ministers habitually allocate to industrial fishing fleets, which fly in the face of the advice and break both international and European Union law. I read books, articles, and reports on marine litter, bycatch, longlines, ghost fishing, beam trawling, ocean acidification, and poor fisheries management. We talked to members of Parliament, civil servants, policy advisers, lawyers, marine biologists, celebrity chefs, and Cornish fishers about destructive fishing techniques, microplastics, illegal quotas, dead zones, and harmful state subsidies. In all that time no one ever mentioned that all of the sea is protected by international law. Didn't they know? I couldn't understand why nobody else had put forward the same solution to all these problems—for governments just to enforce existing laws. Would I be laughed out of the room if I put the suggestion out there? And why isn't the sea safe if the law to protect it is already in place? I decided to find out by looking again at the principal ocean treaty—the Law of the Sea.

Why Doesn't the Law of the Sea Keep Oceans Safe?

The Law of the Sea is a milestone treaty, said to be the most complex and comprehensive international agreement ever concluded. With approximately 180 signatory countries, it clarifies countries' territorial rights and responsibilities, preempting conflict between nations and contributing enormously to global peace. There are only a handful of nonparty countries, including Turkey, Israel, Venezuela, and the United States.

The United States wasn't party to the original agreement because there were objections to part XI, some provisions being judged as disadvantageous to the American economy and as infringements on the country's sovereignty while favoring the economies of Communist nations and some developing states. The treaty was revised to offset these objections and duly signed by President Clinton in July 1994. However, despite the modifications some Republican members of Congress with an entrenched distrust of the United Nations continue to block its ratification. This means that the United States is the only major power not to be a legitimate party to the Law of the Sea.

Although their motives are largely self-serving, many influential sectors of commerce and industry such as oil, energy, mining, and shipping recommend ratification (i.e., giving national legislative consent to be bound by the treaty, also called formal accession). All branches of the American military consider it essential for the easier movement of the armed forces abroad, allowing military and commercial ships navigational rights through foreign seas and international straits. In addition, the US Chamber of Commerce, representing over three million businesses of all sizes and sectors, supports ratification of the Law of the Sea. Russia, Canada, the United States, Norway, and Denmark claim rights to the Arctic's natural resources, and many people wonder if the United States can reasonably stake claims to Arctic oil and minerals without ratifying the treaty. Strictly speaking, the Americans aren't even entitled to a seat at the discussion table.

From the environmental perspective, being a party to the agreement could strengthen the American position in tackling urgent problems such as collapsing fish stocks and ocean acidification. Because the Law of the Sea is generally accepted as the codification of long-standing customary laws, opponents to ratification argue that there is nothing to be gained by formal accession to the treaty because the United States already observes customary international law. (This is ironic when you consider that the United States, along with the other nonsignatory countries, benefits hugely from the international stability the agreement brings, which exists only because of the cooperative efforts of all the countries that do engage and do participate.)

The Law of the Sea has been described as visionary, as a remarkable achievement in global consensus, as a turning point in international lawmaking, and more. It may be all of these, but it is failing to safeguard the marine environment and its wildlife. As it stands, the law is splintered, it can be vague, and it lacks power. It was signed in 1982 and came into force in 1994. There have since been many technological developments that aren't accounted for, such as advances in the ability of fishing vessels to track and capture fish, accelerating the demise of many species. The search for minerals and biological organisms for use in pharmaceuticals and cosmetics, known as bioprospecting, was virtually unknown at the time the Law of the Sea was written. Furthermore, melting ice caps, rising sea levels, ocean acidification, and the plague of plastic debris had yet to make their deadly entrance onto the world stage.

Another major shortcoming is the way the law is implemented (or rather *not* implemented). Instead of having an integrated governance structure, the Law of the Sea is administered by a group of unconnected agencies that separately manage various spheres of activity. There is the International Sea Bed Authority to control undersea mining, the International Maritime Organization to regulate shipping, and the Regional Fisheries Management Organizations (RFMOs) to manage fishing in areas beyond countries' EEZs (known as the high seas). However, the biggest weakness in the Law of the Sea is just that—weakness. Although it includes specific requirements for states to enforce the laws and regulations laid out in the treaty, the condition of the sea and diminishing populations of wild species show that this is not happening. Without a cohesive system of governance and compliance to make sure it is universally enforced, the law is ineffective. It is a guard dog with no teeth.

Members of the drafting committee of the Law of the Sea negotiated for years to reach agreement. If they were wise enough to secure reasonable protections for living marine resources, surely they wouldn't have legislated something that wasn't achievable? It was Committee Three of the Convention that drew up articles concerned with conservation. Perhaps I could ask one of the committee members the question directly. Sadly, it was too long ago. Those I traced had died, but I discovered that Tommy Koh, president of the Third Conference on the Law of the Sea from 1980 to 1982, was still with us and living in Singapore. I emailed Professor Koh to ask how the drafting committee members, who set down safeguards for marine environments, had anticipated their being implemented. I was delighted to receive a reply only a few days later. He explained that the committee had been too optimistic about the people who would be putting the terms of the treaty into practice. In short, they were thought to be more capable and more committed to the spirit of the agreement than they have been, and they were expected to manage the use of marine resources very much better than they do.

The Law of the Sea is not the only international law intended to protect the marine environment, but other multilateral agreements have also fallen short in doing so (such as the Convention on Biodiversity and the World Charter for Nature). I wondered if there is something about international law itself that hinders the success of these agreements.

The Nature of International Law

The subject of international law sounds daunting and complicated, and it is. I wanted to understand why it isn't working for the sea. There are international laws on an array of issues such as trade, human rights, public health protection, air travel, shipping, diplomatic immunity, and the environment. In essence, they are written records of agreements made between countries on rules of engagement and practice, long-term objectives, and common principles—normally called treaties or conventions. The era of globalization brings more complex interaction and cooperation between countries, and with it ever more treaties.

Fortunately, most countries obey most international laws most of the time, rather the way most citizens obey most national laws most of the time. It seems that to a large degree countries' behavior mirrors citizens' behavior. The question is why? A country has government—a judiciary, law enforcers, and penalties—to ensure that people abide by domestic law, but there is no global government with the authority to police and enforce international law. And yet mostly we comply. Clearly there are other motives for being a good, law-abiding country. Understanding them could explain why we don't look after the sea, in spite of the law.

Going back to the citizens parallel, I realized that the threat of prosecution and penalty is not the only reason I obey the law. Mainly that's just what I do and what others expect me to do. It's how we behave in a civilized society. There's an underlying understanding about being in a community: if we live by the rules and pull together, our lives will be safer, happier, and better. And so it follows that if countries behave like people, mostly they will stick to the rules for similar reasons. Once the terms of a treaty have been internalized by the state and adopted into domestic law, compliance usually becomes standard practice. And countries also benefit from more positive interaction with each other, better relations, and easier running of state affairs. Another significant factor is simply doing what you agreed to do and not losing the respect of the global community.

With this backdrop of relative success, where does international law go wrong for oceans? First, generally when a country seriously breaks international law it will provoke condemnation from other parties to the treaty. Diplomacy often resolves the problem, and if it doesn't, collective naming and shaming can work, since no one likes negative public attention. After a while more severe action may be taken, such as imposing trade sanctions and, in extreme cases, even military intervention. In any event, the international community puts some sort of pressure on the errant state to fall in line. But when it comes to the sea the weight of the upright majority isn't available to force compliance because much of the upright majority is itself breaking the laws that protect it.

Second, shared resources lying beyond territorial jurisdictions, like wild fish in the high seas, are in a stateless, unpoliced place and open to overuse by free riders. If they are not sufficiently controlled by the flag state (the country where a ship is registered) and are acting only in their own interest, commercial fishing fleets tend to overexploit the commons pool resource—the ocean—and waters become too heavily fished. In his influential article "The Tragedy of the Commons," Garrett Hardin described the cycle of degradation of resources in open access areas using the analogy of sheep overgrazing common land.[6]

And third, recognizing the rewards of keeping international agreements is a key factor in compliance. Wider society has not understood how much there is to lose by not safeguarding the sea and marine life. Most of us simply don't see the connection between healthy oceans and a better future (in fact, any future). To put yourself out, to make the effort, to invest time and money to protect something from harm, you must value it or love it. Probably the biggest problem of all is simply that we don't care enough. Recognizing this is essential to winning the battle for the oceans.

There are exceptions, but when it comes to the living resources of the sea, commonly the majority of national governments appear to break the law routinely and allow industry to do the same. Consequently, conservationists are battling to create protected areas in places where waters and wildlife are lawfully protected anyway. All life on Earth is ultimately dependent on healthy and life-rich oceans; therefore, without intelligent, cooperative management of our use of the sea and with no effective mechanism to enforce the law, in the words of the wise—we're stuffed.

It was late October; the weather had broken and it was chilly and damp. After plenty of reading and listening and thinking, it was time to write things down. I had begun by asking myself why we are struggling to safeguard patches of the world's oceans and leaving the rest still exposed to serious damage, depletion, and pollution. When you consider that oceans are the planet's life support, this makes no sense. Even allowing for the dominance of today's money-obsessed and profit-driven culture, it still makes no sense. However wealthy people are, they can't do without a survivable climate and clean air to breathe, nor can their children and grandchildren. To compound the illogicality, I then find that the world's oceans are protected by laws that are mostly ignored.

My weary laptop flickered to life. A blank document appeared and waited for words: empty, white, defiant. I grabbed a coat and went for a long walk along the river, undeterred by the autumn drizzle. The path was empty of walkers and broken with puddles. As I put my mind in neutral and kept going, I could already hear the opponents' voices in my head. "Industry would never allow it." "But how? How could it happen?" I was also aware that our use and misuse of marine resources is not the only cause of the troubles at sea. Even if the whole marine world were protected as the law says it should be, that still wouldn't make everything right. There are serious threats to the well-being of coasts and oceans caused by what happens on land, probably the greatest being too much carbon dioxide released by our fossil-fueled lives and the consequent effects of climate change. Although shipping accounts for a proportion of marine pollution, most is from land-based sources like sewage, industrial discharge, and agriculture's chemical runoff into rivers and the sea. Nor could we forget the dreaded plague of plastic waste, washed in from landfill and through drainage systems, items large, small, and microscopic—left on beaches, floating on the swell, lingering in the water column, and taking over the seafloor.

Pretty soon I was wet and cold. I looked into the mud and began thinking of cake. Things became clearer. Could I build a convincing case for protecting the entire global ocean and gain support for it? I turned around and walked home, the rain easing as I got to the door. I threw off my coat and ran upstairs, going straight to the laptop and the expectant blank page. The answer to the question, "But how?" could come later. First comes the most powerful part of the case for change—to consider just what we are protecting.

3: Teeming Seas

I found it funny to hear a microbiologist on the radio getting excited about evidence that a single-celled organism might have existed on Mars thousands of years ago when here on Earth there are so many forms of life that no one knows the number—and they're living now. The diversity of life on land is astonishing enough; in the ocean it's unimaginable. Some biologists say there are about a million marine species, and others believe there are up to ten million. The range of the estimates alone reveals how little we know about the world under the waves.

On land, livable space constitutes a fairly thin layer of possibility; in the sea it has a far greater dimension. Space for life reaches from the surface through the water column down to the seafloor, which may be tens of thousands of feet below. The average depth of the global ocean is about 2.2 miles, and at its deepest, in the Mariana Trench in the Pacific, it's almost 6.8 miles, which means that about 98 percent of the living space on Earth is on the waves or beneath them, in a saltwater kingdom. In part this explains the untold variety of ocean life.

In June 2016 the headline of a report on the Smithsonian's website read, "Mission to Mariana Trench Records Dozens of Crazy Deep Sea Creatures." For three months scientists on the *Okeanos Explorer*, the research ship of the US National Oceanic and Atmospheric Administration, mapped the seafloor and recorded marine life in the great depths of the ocean, using a remotely operated underwater vehicle. They found previously unknown species; jellyfish, octopuses, worms, and comb jellies, plus an anemone with tentacles six feet long. "Every time we make a dive we see something new. It's mind-boggling," said Patricia Fryer at the University of Hawaii.[1]

Generations of biologists and taxonomists have classified each newly discovered life form and found its place in a sprawling index of descending sections and subsections. Corkscrew anemones, for example, are members of the animal kingdom, phylum Cnidaria, class Anthozoa, subclass Hexacorallia, order Actinaria, family Hormathiidae, and genus *Bartholomea*. There are over 1,000 known species of anemones, about 5,000 species of sponges, 9,000 of segmented worms, over 45,000 of crustaceans, and more than 110,000 recorded species of mollusks. (There is debate among scientists about which category some organisms belong to.)

Among the ocean's sublime life forms are phytoplankton—microscopic free-floating plants drifting throughout the world's rivers and seas. They make their food through photosynthesis, using the energy of sunlight to convert carbon dioxide and water into glucose, leaving oxygen as a by-product. Phytoplankton take bizarre forms and fantastic shapes. One of 40,000 known types, *Odontella aurita*, for example, looks like trembling slices of light flecked with scraps of amber and quartz and linked by threads of gold. But it is remarkable less for its beauty than for helping to run the engine of life on Earth. As well as being the foundation of a complex food chain that feeds the ocean, NASA estimates that, each day, over 100 million tons of carbon dioxide is drawn from the atmosphere into the ocean by phytoplankton. And together with seaweeds and sea grasses, it is thought that phytoplankton release well over half of the Earth's oxygen into the atmosphere (more than rain forests and other land plants combined). When I first found out about these miraculous little plants I needed a few moments to take it in. I stared blankly out the window for a while, picturing them swaying in the water in their billions, busily making the air we breathe and providing food for the fish we eat.

Ocean life ranges from bacteria just thousandths of a millimeter wide to a blue whale 108 feet long and weighing 220 tons; from a frigate bird soaring two and a half miles above the sea surface to an eerie snailfish living over five miles below it. We are bewitched and bewildered by the variety, the beauty, the drama, and sometimes the plain weirdness of life in the sea. For most people the most interesting examples of sea life are fish, whether to catch, to sell, to cook, to eat, or just to watch. And the massive demand for them leads to one of the most acute threats to the global ocean: commercial overfishing— taking too many wild fish and using methods that ruin habitats and kill great numbers of nontarget species as bycatch. It also causes millions of sea creatures to starve when humans take their food supply. Therefore fishing responsibly, with moderation and good sense, will put right a great many of the troubles at sea. It is absolutely possible to do that and still have plenty of fish in the sea—far more than there are now.

Approximately 31,500 species of marine fish have been identified, from a tiny Indonesian *Paedocypris*, smaller than your fingertip at 0.31 inches, to a whale shark as long as a bus, averaging 32 feet at maturity, with some reaching over 40 feet. Fish come in every conceivable color and in countless combinations. They are striped, spotted, checked, marbled, and dappled. They are two-toned, multicolored, transparent, iridescent, and luminescent. There are flat fish, round fish, egg-shaped, wedge-shaped, and even cube-shaped fish, such as the wonderful yellow boxfish. Fish live in estuaries, in rock pools, among mangroves, in deep-sea crevasses, on tropical reefs, in open oceans, and in arctic depths. Some species use camouflage for self-defense or to dupe their prey, pretending to be coral, sea urchins, rocks, pebbles, sand, or kelp. I once was fascinated by a leafy sea dragon at the aquarium in Barcelona, rippling in midwater, mimicking seaweed, festooned with ribbons of green and yellow like a seahorse at a fancy dress party.

Fish propel themselves in various ways: undulation, oscillation, and dynamic lift. Some skulk a few yards above the seafloor or flit gracefully over the coral, while sailfish and marlins charge through the ocean at up to sixty miles an hour. There are loners, such as John Dories and sunfish, and there are the sociable and insecure, like sardines and herring. Even today schools of Atlantic herring, estimated at up to four billion individuals, swim great distances on a triangular route between spawning, feeding, and nursery grounds. This equates to over a cubic mile of cruising fish, and their numbers were even greater in the past. In *A History of the Earth and Animated Nature*, published in 1774, Oliver Goldsmith writes,

> The whole water seems alive, and is seen so black with them to a great distance, that the number seems inexhaustible. There the porpoise and the shark continue their depredations, and the birds devour what quantities they please. By these enemies the herring are cooped up into so close a body, that a shovel, or any hollow vessel, put into the water, takes them up without farther trouble.

There are other such historical accounts from across the world, conveying the abundance of marine life there once was—an extraordinary abundance compared with today. The comparison reveals how much we have lost and shows how far the oceans' wild populations would have to rebound to be truly as nature intended. With a gradual decline of ocean plenty, each generation has a reduced notion of what "plenty" means. When I was a child and we went rock pool fishing on family holidays, my father would have two gripes: that there were fewer crabs and fish to catch than when he was a boy, *and* that they were smaller. I was happy, though. For me the amount and size of sea life in the little pools was normal. Years later, when rock pool fishing with my own children and finding even fewer crabs and fish, I was the one bemoaning the diminished number and size of rock pool life "these days," but the kids seemed quite content with what was there. They didn't know what should have been in the pool, nor did I or my father. You would have to travel back several hundred years to see how much life there could be in a rock pool on a southern English shore when few humans were around.

In his article "Anecdotes and the Shifting Baseline Syndrome of Fisheries," scientist and marine biologist Daniel Pauly put a name to the diminished view of what is regarded as "normal":

> Each generation of fisheries scientists accepts as a baseline the stock size and species composition that occurred at the beginning of their careers, and uses this to evaluate changes. When the next generation starts its career, the stocks have further declined, but it is the stocks at

that time that serve as a new baseline. The result obviously is a gradual shift of the baseline, a gradual accommodation of the creeping disappearance of resource species.[2]

The shifting baseline syndrome blinkers and misguides us. When fisheries managers recognize this problem of perception and take it into account, they make wiser regulatory decisions, such as setting lower catch quotas and closing more nursery areas to enable fish populations to rebuild and habitats to recover (even if they could probably never return to their untouched, pristine states). In 2017 the British press was reporting that North Sea cod were "back" after a spell of draconian fishing restrictions stopped them from edging further toward extinction. Catch quotas could increase now, and yippee, everything was fine. Cod were back in the net and back on our plates. But the relative baseline for cod being "back" was set at an already overfished stock—less than 10 percent of what it was two hundred years earlier, before industrialized fishing began.[3] That isn't "back," it's "coming back a little." Only two years later, North Sea cod were reported to be at critically low levels again. Phil Taylor from the conservation organization Open Seas explains why: "Seeking short-term profits, ministers have undone or are undoing many of the recovery measures. Last year quotas for cod in the North Sea were set twenty per cent above the scientific advice."[4]

The Limitless Larder

For centuries herring were vital for the people of northern Europe, particularly the poor. They were a plentiful and affordable source of protein, and thousands of jobs depended on these silver darlings, as they are often called. There was work catching them, smoking or salting them, transporting them, and selling them. Herring are very important to marine ecosystems. Feeding mainly on tiny zooplankton, they are a critical link in the food chain, converting plankton into thousands of tons of fish flesh for the mouths of predatory creatures farther up the chain. They are a principal source of food for a wealth of wild species such as puffins, gannets, razorbills, arctic terns, and several types of gulls, as well as for sharks, cod, tuna, bass, mackerel, and salmon. The attention of predators can trigger panic in the schooling herring. Desperate not to be eaten, each one tries to hide in the center of the school, creating a great whirlpool of shimmering bodies known as a bait ball. Mammals feed on herring too: dolphins, porpoises, whales, seals, and of course humans, who frequently take more than their share. In the 1970s the entire North Sea herring fishery collapsed. Stocks recovered to a degree, but by the mid-1990s further overfishing led to another collapse.

Thomas Huxley, a much respected English biologist, was a public defender of Charles Darwin and his theory of evolution. Speaking at the International Fisheries Exhibition in London in 1883 he announced, "I believe that the cod fishery, the herring fishery, the pilchard fishery, the mackerel fishery, and probably all the great sea fisheries, are inexhaustible; that is to say, that nothing we do seriously affects the number of the fish. And any attempt to regulate these fisheries seems consequently, from the nature of the case, to be useless." Huxley's thinking encapsulates perfectly why commercial fisheries have too often been managed badly. If you begin with the belief that stocks are boundless and undersea habitats are indestructible, things have to go wrong eventually. A striking photograph taken in the English port of Grimsby in 1914, about thirty years after his speech, helps explain why Huxley would make such a bold and misguided statement. Enormous heaps of enormous fish cover the quayside, with men unloading more from trawlers moored in the background. For much of the twentieth century, Grimsby vied for the title of biggest fishing port in the world with the city of Kingston upon Hull (also known as Hull), which sits on the opposite bank of the Humber estuary. Up to thirteen trains a day loaded just with fish were leaving Grimsby for the markets, shops, hotels, and restaurants of towns and cities across the land. British waters and the distant fishing grounds of the North Atlantic seemed to be infinitely productive, and for centuries they gave generously. Among the species served up on the nation's plates or wrapped in newspaper with chips were cod, herring, hake, haddock, mackerel, and common skates (now critically endangered in many areas).

Huxley was wrong, of course. He could not imagine the industrial-scale fishing to come, nor could he foresee that even the most plentiful waters cannot keep up with relentless taking year after year, decade after decade. Fisheries management needs to be based on facts, not speculation. Nothing illustrates this better than the collapse in 1992 of the Grand Banks cod fishery off the coast of Newfoundland in Canada, a notorious cautionary tale of disastrous fisheries management. For centuries the area was regarded as having the best fishing grounds in the world, attracting ships from hundreds of miles away: "Their huge nets took unprecedented amounts of fish, which they would quickly process and deep-freeze. The trawlers worked around the clock, in all but the very worst weather. In an hour they would haul up to 200 tonnes of fish; twice the amount a typical 16th century ship would catch in an entire season."[5] Despite warnings from scientists and inshore fishers, the Canadian authorities ignored the evidence and would not reduce catch limits. In time the cod population completely collapsed, the fishery was closed, and about forty-five thousand jobs were lost.

As long as fishing remained small-scale and nonmechanized, the limitation on catchable amounts kept fish populations reasonably safe. Then came the industrialization of fishing; larger ships with mechanical winches could pull massive nets, and new types of gear found ever more ways to catch the fish, plus variations of trawling the seafloor, leaving a trail of unseen devastation. Steam and then diesel power replaced wind and sail, allowing ships to go farther, into previously unfished waters. Refrigeration meant that catches could be stored for weeks, so ships could stay out fishing longer, catching more fish. Finally there came sonar and satellite technology to locate and track fish even deep beneath the waves. Large-scale fishing has become like a military operation, and these days the chances of target fish escaping capture are minimal.

While the dark clouds of climate change gather on the horizon and the plastic pandemic becomes ever more pervasive, the most destructive human activity happening in the ocean today is industrial fishing, and yet arguably it is the most straightforward issue to deal with and the easiest to prevent.

The report of the Food and Agriculture Organization (FAO), "The State of World Fisheries and Aquaculture" (2018), found there are approximately 4.6 million fishing vessels operating in the world and that almost two-thirds of commercial stocks are being fished to capacity, with a further third fished at biologically unsustainable levels. Drawing from a range of information sources not used in the official data, marine biologists Daniel Pauly and Dirk Zeller published research showing that the true level of global fishing is up to 50 percent greater than the FAO's figures.[6] Illegal, unreported, and unregulated fishing (IUU) seriously compounds the problem. IUU is fishing out of season or using prohibited gear and banned practices, or it could be fishing in a protected area, catching a protected species, or catching over the legal quota. It could also be operating without a license to fish, or flying a "flag of convenience" to escape inspection (when a ship is registered in a country with minimal regulations), or using forced labor or child labor on board. It is very difficult to know how much IUU accounts for, but estimates are between 20 and 30 percent of the world's catch, with an approximate annual loss of US$23 billion.[7]

Fisheries management is a hefty subject, and as I learned about the successes and failures of different methods one truth stood out: people are trying to manage the capture of a living natural resource that we cannot see or count, that is forever moving about, and that we don't sufficiently understand. Therefore it is not surprising that finding a fail-safe method of control is not easy. Imagine running a grocery store where you aren't sure how much stock there is or exactly what the stock is, where the stock won't stay still and all the lights in the storeroom are out. Nevertheless, with the right management mind-set and enough political will, overfishing and destructive fishing could end tomorrow. We know what causes it, we know how to stop it, and we know what to replace it with. The depletion of fisheries around the world shows that long-standing, conventional management methods aren't working, and if they carry on we will be left with seas that are nearly fishless. Their scope is generally narrow, profit-driven, and focused on single-sector interests and single-species catches. The older, well-established methods barely account for the complexities of natural systems that give rise to the target species in the first place: the physical, chemical, and biological factors. For instance, if the primary food source of a commercial species is another type of fish, how are those stocks affected? Other fish, seabirds, and marine mammals will also depend on forage species. As the Lenfest Ocean Program reports,

> A task force of scientists, for example, recently described the major impact of little fish such as herring, anchovies, menhaden, and sardines and concluded they have twice as much value left in the water to feed other fish than as target catch themselves, even though they account for more than one-third of wild marine fish caught globally. How managers allow for the role of these forage fish is just one of the remaining frontiers in our understanding of ocean ecosystem.[8]

In other words, do we prefer cod or herring, mackerel or sardines? Do we want the predator or the prey? We can hardly have a wealth of both if things don't change.

"Bad" sea fishing is a matter of not only the volume of fish taken, but also the very destructive ways they are caught. These methods include bottom trawling (dragging trawls across the seabed), cyanide[9] and dynamite fishing,[10] muro ami,[11] shark finning,[12] longlining,[13] and indiscriminate fishing (where giant nets catch everything including nontarget species). Add to this the absurdity of subsidies, with several countries giving billions of dollars of the public's money to support fishing industries. A portion of it is spent in beneficial ways, perhaps on small-scale fishery development or on monitoring and enforcement, but well over half of state subsidies helps drive overfishing and perpetuate harmful fishing practices while propping up companies that are not commercially viable. The lion's share of that funding goes to industrial fleets, largely toward the cost of fuel (especially benefiting distant fleets operating in the high seas). Payments also offset other costs, such as building vessels and upgrading fishing gear. The payments keep the global fishing fleet much too large—two and half times the capacity compatible with available stocks.[14] The European Union, for instance, props up an unprofitable and unsustainable fishing fleet with over €1 billion of taxpayers' money each year.[15] There's a new old adage, "Give a man a fish and he eats for a day; subsidize a man to fish and he'll empty the ocean and trash the seafloor."

Christopher Pala has written,

> Not only are most industrialised ships uneconomical if they aren't subsidised, they also provide far fewer jobs: 200 for every 1,000 tonnes of fish caught, versus 2,400 jobs for 1,000 tonnes caught with artisanal methods using small boats. . . . [N]early all of the fish the small-scale fishers catch is eaten, while the industrial ships, in addition to the thirty million tonnes of edible fish they take, also haul out another thirty-five million tonnes of everything from other fish to plankton for transformation into oils or fish meal, used for fertiliser and feed. The result: many of the non-food fish that the edible fish depend on have disappeared, along with vast amounts of plankton, the base of the food chain.[16]

Picture a subsidy turnaround, with large sums of state funding going toward conserving marine resources instead of exhausting them. The money could be helping the fishing industry adopt better methods of working—to keep undersea habitats intact and commercial stocks buoyant. There are hopeful signs that change is coming to the iniquitous world of fisheries subsidies. For several years, World Trade Organization negotiations have been under way to address the issue, and extra pressure to reach agreement came with the United Nations Sustainable Development Goal for oceans, Life Below Water: "By 2020, to prohibit certain forms of fisheries subsidies which contribute to overcapacity and overfishing, eliminate subsidies that contribute to illegal, unreported and unregulated fishing and refrain from introducing new such subsidies."

When you read about badly managed fisheries, it is clear what well-managed ones should be like. The aim is to have resilient seas brimming with life, maintaining the natural balance and interrelations between species, with plentiful fish stocks to provide food and employment in perpetuity. But how do we achieve that? How do you banish the scourge of ruinous commercial fishing? In a sense it's easy.

You find places where they fish in a different way, in a better way—
then you do the same.

Getting Back Fish-Filled Seas

Despite the apparently dismal picture worldwide, there are places
where good fisheries management has replaced bad. If the rest followed their lead, the formula for responsible fisheries policies could
eventually wrap around the globe.

In towns and villages along the coasts of the United States, the
words "Magnuson and Stevens" can evoke a quiet reverence and heart-
felt praise from the most hardened and antiestablishment of sea-
faring characters. What they fondly call "Magnuson" is a law named
after two senators instrumental in its development and passage in
1976. The Magnuson-Stevens Fishery Conservation and Management
Act (MSA) completely restructured the way American fisheries were
regulated and brought them back from the brink of collapse, rebuilding depleted stocks and protecting a valuable resource for present
and future generations.

Gary Jarvis is a burly yet soft-spoken fisherman from Destin,
Florida. Once a fishing village, the town has become a major tourist
resort. People come for the pearly white beaches, turquoise waters,
and recreational fishing in the Gulf of Mexico. Gary is one of many
people in Destin who rely on healthy local fish stocks to make a living.
Standing at the helm of his cruiser moored in the harbor, he talks
about how the Magnuson-Stevens Act turned things around for the
community, and not only for Destin:

The story you can tell in Destin, you can tell in Pensacola, or Browns-
ville, Texas, or Rhode Island or New England—fishers and fishing
communities have a common bond. We all have the same concerns,
the same challenges, trying to be businessmen, to raise our families
and provide for them, and do it off this resource. Dealing with Mother
Nature on a daily basis, it's a tough life and it was hard to prosper as a
fisherman, and some of it was self-inflicted. Before Magnuson came

along, a lot of our fisheries were depleted, and we did it to ourselves. Without Magnuson, we would kill the resource.[17]

The MSA created eight regional fishery management councils calling for representation from all interested sectors in planning and management. Scientists, fishers, managers, conservationists, and citizens work together and determine appropriate catch limits that allow fish populations to recover and keep them at healthy levels.

Another Destin fisherman, Jim Green, welcomes the MSA's inclusive and participatory style of management:

> By allowing these protections to be put in place, by having a stakeholder-driven process where we can go to a regional council and give our input into the things that are good, the things that are bad and the things that need to be changed has really opened the door for us. Magnuson is about all of us working together, tailoring our management to each individual sector's needs and creating a thriving fishery.[18]

The American regime is not perfect by any means. The Pacific Fishery Management Council, for instance, has been criticized for setting catch limits too high for the anchovy and sardine fisheries off the West Coast and for contributing to the collapse of stocks that are an important forage food for a host of other species. Consequently, there is some way to go in the United States, but they have made the all-important break with the old system and a myopic belief that we can constantly cheat nature and focus only on short-term profit.

Clutching the western edge of northern Europe, Norway is known for its spectacular fjord-riven scenery and for having some of the best-managed fisheries in the world. The serrated coastline is the longest in Europe and reaches well beyond the Arctic Circle. Norway's waters stretch westward into the North Atlantic and into the Barents Sea, making its exclusive economic zone roughly four times the land area, so fishing potential is immense. These waters are rich. As well as ample stocks of fish, there are large populations of birds and marine mammals such as orcas, fin whales, and humpback whales. Commercially caught species include herring, cod, capelin, mackerel, pollock, blue whiting, haddock, and prawns, and fishing has long been the lifeblood of communities along the coast.

Most commercial fish species are transboundary and managed in cooperation with other countries, and Norway has fishery agreements with the European Union, Russia, the Faroe Islands, Iceland, Greenland, Poland, and Sweden. For instance, Norway and Russia share fisheries management in the Barents Sea, which reaches up from their northern coasts to the Arctic Ocean. The Barents cod stock is the world's largest and bears out the benefit of nations' working together. With a strict science-based quota system in place, it also demonstrates that, for commercial fishing, by taking less we gain much more. With fisheries management, politics also count, especially in Norway. Having chosen not to be a member of the European Union, the Norwegians can control their fisheries independent of EU regulations. The result is one of the world's best fishery regimes running alongside possibly one of the world's worst: the EU's Common Fisheries Policy.

So how do Norway and the United States do it? Both regimes are founded on the principle that living resources have limits and that fisheries have to be managed for the present and for the future. Much of the strategy is about keeping plenty of breeding fish in the sea to maintain stocks. These are the big old fat and fecund female fish, called BOFFFFs. They have the most eggs and the biggest ones. For example, if left to mature, an average adult female Atlantic cod would lay from four to six million eggs in a single spawning. The bigger she grows the more eggs she will produce, and these fish can reach six feet and weigh over two hundred pounds, although cod this size are rare nowadays. Actions the two regimes have in common are having catch limits set according to scientists' advice and not by politicians; applying robust monitoring and enforcement measures throughout the industry; using innovative gear and technology to reduce bycatch and discards; involving a wide range of stakeholders in planning; and closing biodiversity-rich areas seasonally or permanently—all braced by supportive legislation. Another significant factor in Norway's and America's success is a drive to understand more about natural systems and to constantly improve the way things are done (known as adaptive management), unlike the static "business as usual" approach of many regulating authorities.

Although it is commonly considered a model of sustainable management, Norway's method has its critics too. They ask, Can a fisheries policy that includes whaling be called well managed? Despite opposition from several organizations, including the International Whaling Commission and Greenpeace, Norway still practices commercial whaling. On average, 550 to 600 minke whales are killed each year. Authorities give three principal reasons for continuing to hunt minkes. One is the supposed cultural right to do so, known as tradition, yet according to Greenpeace minke whaling didn't begin in Norway until 1930. (Besides, isn't labeling something a "tradition" often a pretext for continuing a bad habit?) A second reason given is public demand for whale meat, which they say is commonly part of the Norwegian diet. In fact, people have little appetite for it. "Fewer than 5 percent of Norwegians regularly eat whale meat, and those who do, consume on average about half a pound each year. To stimulate demand, the Norwegian government has subsidized the business and funded marketing initiatives," says Kate O'Connell, a marine wildlife consultant from the Animal Welfare Institute.

A report by the Environmental Investigation Agency and the AWI found so few people like the meat that much of it ends up as pet food or feed for foxes on fur farms. In 2014 more than 125 tons of whale meat (equivalent to the marketable meat from seventy-five minke whales) was delivered to Rogaland Pelsdyrfôrlag, the largest manufacturer of animal feed for the Norwegian fur industry.[19] As Jennifer Lonsdale, cofounder and director of the EIA, notes, "The Norwegian government claims it is important to have whale meat as a source of food for people, but because of falling demand, the product is now being exported. Now we discover it's going to feed animals in the fur industry, which we find completely unacceptable."[20]

Last, it is claimed that minke whales are eating too many fish (as if they have a choice) and are therefore detrimental to the fishing industry and must be culled to protect fish stocks. Experts disagree. Whales play a crucial part in maintaining healthy underwater ecosystems. A study published in *Frontiers in Ecology and the Environment*, titled "Whales as Ecosystem Engineers,"[21] concluded that the presence of whales gives rise to more fish in the sea, not fewer. As they dive to feed, then return to the surface to breathe, whales stir up nutrients and microorganisms from the seabed right through the water column, creating an underwater feast for other sea life. Whale urine and excrement also make excellent fertilizers and so boost the production of phytoplankton. Deaths are useful too. The decaying carcass of a whale feeds a unique range of scavengers on the seafloor. "Although Norway has a reputation for progressive environmental policies, its credibility is undermined by its whaling policy."[22]

The Norwegian and American regimes are works in progress. If we think of the move away from failed conventional fisheries management as a journey and the ideal system as the eventual destination, they are well on their way and there is a great deal to take away from the policies of both countries. They confirm that with common-sense stewardship, fish and wildlife populations bounce back with a vengeance, bringing the added benefits of more food, more jobs, and more prosperity. Legislators, policy advisers, and fishery managers of other regimes—those in a position to force change—have only to see what is already working and do the same.

Fish-Filled Seas and More

Getting the fish back is one part of managing the use of marine resources well, but there's more to it than just making commercial fisheries bountiful so as to be bankable. Sustainable and well managed aren't necessarily the same thing. It depends on what the management is good for. The environment? Wildlife? Or profits and exports? The true value of life can't be measured in dollars and kroner, euros and yuan alone. Conservation biologists now favor rebuilding all species populations and safeguarding biodiversity as the best goal of fisheries management, rather than just planning for bumper commercial stocks. The purpose of ecosystem-based fisheries management is to consider the whole ecosystem and its wildlife, recognizing the complexity of food webs and the interdependence of commercial and noncommercial species. Pitcher and Pauly note that "sustain-ability is a deceptive goal because human harvesting of fish leads to a progressive simplification of ecosystems in favour of smaller, high turnover, lower trophic level fish species that are adapted to with-stand disturbance and habitat degradation."[23]

Norway and the United States have a hybrid approach because their fisheries regimes are in transition—rooted in the old style while embracing elements of a quite different and more progressive way of doing things. By 1990, people from fishing communities, scientists, and conservationists were talking about a new way to manage not just fishing, but everything humans have to do with the sea, because fishing responsibly is only one step in repairing damaged oceans. This new type of management is known as the ecosystem approach (EA) or ecosystem-based management.

Picture a common scenario where the needs of the people, the de-mands of industry, and the pressures on the environment clash. Per-haps an aggregate company wants to dredge an area of the seabed, or trawling is ruining the nursery grounds of certain fish species, or fertilizer runoff from nearby farms is polluting inshore waters and local businesses are losing custom because visitors are going to places where the sea is cleaner. All the parties involved (called stakeholders) come together. They may be fishers, farmers, conser-vationists, business operators, and representatives from govern-ment departments and from industry. By discussing how their ac-tivities affect the sea and each other, and by listening to everybody's concerns, they are able to build a big-picture understanding of the situation. "As with many of these processes, the act of going through the consultative process to develop the plan is just as important as the output itself. It engenders ownership of the plan, trust of other stakeholders and starts to build up a sound working relationship be-tween players."[24]

Planning should be underpinned by sound scientific advice. The overall objective is to safeguard natural resources (marine, terres-trial, and atmospheric) for the benefit of all. People are also a part of nature's machinery, so the ecosystem approach allows for the social and economic demands of us land-living creatures as well. When a cohesive and coordinated management plan is agreed to and put into effect, a spirit of cooperation develops. Progress is monitored, evaluated, and improved over time. With a clearer understanding of the connections between people and nature and by including all interested parties in decision making, the scheme is more likely to succeed. Of course this is a simplistic illustration of the ecosystem approach, but it conveys the underlying philosophy.

There are varied definitions of it, but in reading or hearing about the ecosystem-based approach certain words and phrases crop up again and again, such as protecting biodiversity, participation, cooperation, equitable benefit, responsible use of natural resources, education, and decentralization. Not wishing to be waylaid by the details of the debate, I'm keeping a broad view for now and summing up EA as cooperative management for the common good that allows marine environments to recover and wild populations to thrive, keeping them continuously healthy and resilient.

The term ecosystem approach emerged in the 1980s and was formalized at the Earth Summit in Rio de Janeiro in 1992. Realizing the value of a more holistic course of action, organizers embedded it as a fundamental principle of the Convention on Biological Diversity. Since then, the ecosystem approach has been incorporated into other multinational laws and agreements, including the United Nations Fish Stocks Agreement and the European Union's Marine Strategy Framework Directive. It is endorsed and promoted by conservation bodies and international agencies such as the UN Environment Programme and the Food and Agriculture Organization of the UN (in its Code of Conduct for Responsible Fisheries).[25] Countries such as Australia, Chile, Norway, Namibia, and the United States have incorporated the EA into domestic legislation relating to marine living resources. And yet, despite its being a widely accepted and highly regarded strategy, long legitimized in law, so far not nearly enough of it is happening on the ground or in the water. The old methods still prevail. It's difficult to dislodge entrenched habits even when they don't serve us or the sea. Consequently, in the world of fisheries policy, if it's happening at all, replacing the outworn approach with a forward-thinking one tends to be a prolonged and incremental conversion —

more a sedate metamorphosis than an overnight reform. The highly valued ecosystem approach is not much use if it remains just a good idea trapped in policy documents and statute books on the shelves of government departments. It needs to escape, along with its much-overlooked companion the precautionary approach.

Are there places, though, where practice does match theory and ecosystem-based management is properly up and running? And if so, does it do the trick? Does it restore and revitalize the sea?

Withdrawing Industry's Assumed Right to Ransack

Lapping at the edge of the world, surrounding the continent of Antarctica, the Southern Ocean reaches up to the Antarctic Convergence, where cold water meets the warmer water of the north. With near-freezing seas, forever whipped by the world's strongest average winds and with no intervening landmass to temper their force, storms are frequent and often ferocious. Towering waves over sixty feet high collide, breaking in giant crests and steep valleys of thunderous water. Here nature is at its harshest and most unforgiving, as though expressing the fury it feels. Yet despite the wild and inhospitable conditions, people still come here to profit from the bountiful polar seas. They came first for the seals in the late eighteenth century, then for the whales in the early twentieth, and now mainly for krill and toothfish. Like the orange roughy described in chapter 1, the Patagonian and Antarctic toothfish are particularly vulnerable to overfishing because they are a slow-growing deep-sea species with a high market value. (Once overfished, populations take many years or even decades to recover—if they are left to do so.)

The Antarctic Treaty, initially agreed to by fourteen nations in 1959, secured the continent itself as a place to be used only for peaceful activities and cooperative scientific research, but the ocean around it remained open to commercial exploitation. Perhaps inevitably, by the 1970s fears were growing about the consequences of overfishing, particularly on populations of Antarctic krill, which were being caught in large quantities (for, among other uses, health supplements and fish-farm feed) and in 1982 the krill catch peaked at 581,000 tons.[26] In the Southern Ocean, the biomass of Antarctic krill (*Euphausia superba*) is estimated to be about 417 million tons,[27] greater than the global human population. As krill make their vertical migrations from the surface each night to deeper waters during the day, animals throughout the water column feed on them. Each year more than half the krill are eaten by all kinds of sea life, among them whales, orcas, penguins, seals, albatross and other seabirds, squid, and fish. These shrimp-like crustaceans feed on phytoplankton (and to some degree on zooplankton), making them a vital link in the food chain and a primary food source for marine life. Krill are therefore a keystone species in the Southern Ocean, and without enough of them ecosystems would collapse, leading to the loss of dependent species in a succession of starving wildlife.

Concerned countries came together to find a way to deal with the growing environmental crisis and avert disaster. These discussions culminated in an agreement as part of the Antarctic Treaty, to set up an organization to regulate fisheries and safeguard the Southern Ocean and its wildlife. Consequently, in 1982, with twenty-four member states plus the European Union, the Commission for the Conservation of Atlantic Marine Living Resources (CCAMLR) came into being, and it has since become the most effective body managing fishing and protecting biodiversity in the high seas. Ample stocks of commercial fish species have been secured and are no longer on the brink of collapse. More than a fisheries management body, CCAMLR has the explicit objective of conserving marine living resources by embodying article II of the treaty at its core: "The objective of this Convention is the conservation of Antarctic marine living resources." This means that industries such as commercial fishing can no longer claim they have the right to exploit the resources of the Southern Ocean: they must obtain permission to do so. This is the precautionary approach in action. Putting earned permission in place of assumed rights is central to using natural resources more wisely, whether of the sea or of the land.

What Are the Secrets of CCAMLR's Success?

There are no secrets to using the sea well—it's about being smarter and dropping bad habits. CCAMLR's aim is to find the right balance between people's desire to fish and conserving marine living resources in the long term. In striving to achieve that, the commission has been at the forefront of developing the ecosystem approach to marine management, which allows only rational and responsible harvesting of the sea. A precautionary approach is a critical safeguard when there isn't full knowledge of fish stock sizes in the Southern Ocean or a complete understanding of how its complex natural systems interact. All the commission's policies are based firmly on available scientific knowledge and advice drawn from an extensive database gathered over several decades by various research organizations. One such is the British Antarctic Survey, which has a long history of scientific study in the region. It provides ecological and biological data to CCAMLR, enabling the Scientific Committee to make the best possible decisions about management and setting rules for conservation, such as the appropriate catch limits of target species or new ways to reduce accidental bycatch. It seemed I had found what I'd been looking for: the ecosystem approach in action—a system of controlling what we do in the ocean that keeps ocean habitats safe and defends wildlife populations while allowing people to continue fishing.

Politically, the ocean world is split in two: the water within the territory of a coastal state and the water beyond it—the high seas. Advances in fisheries management mainly affect seas within a country's two-hundred-mile exclusive economic zone because this is where most commercial fishing goes on. These waters tend to have the most fish, and they are the easiest to reach. National authorities also regulate other industries within this area, like dredging, mining, and oil extraction. With decreasing fish stocks in inshore waters, commercial operations are moving to more distant fishing grounds, making the high seas ever more vulnerable to irresponsible or destructive fishing. The extractors and exploiters of nature's bounty go far and wide, and many will take what they can, however they can and hang the consequences. CCAMLR has shown that damaging industrial fishing can be virtually brought to an end beyond national jurisdictions, out in the open ocean, but is it possible to do that over the rest of the high seas—the greatest wilderness on Earth?

4: The Free Sea

For four days whole we endured violent gales from the south east and were thrown all ways by mountainous waves. I could barely breathe from fear and surely could not write. On the morning of the fifth, indeed the 27th day of March, our wind hauled around northward and waned mercifully setting the waves to ease. The sea became silver, near flat as stone while the jib and foresail began to wilt, hoping for air. Great numbers of whales of all sizes arrived beside us at starboard, less than one eighth of a league distant, riding the swell effortlessly and sending up white plumes of breath toward the sky. Our position is 31 degrees south, 116 degrees west and with one hundred and seventy fathoms of line we cannot reach the sea bed. With little joy I look over this monstrous carapace of water to all sides of our solitary ship. I pray to our Lord and to our blessed Mary we are not be-calmed in this savage, wet desert of the high seas. I keep faith in the great timbers, canvas and the rest that went to make this vessel. Be they sound and sturdy and with the strength of our men deliver us safely from this unearthly place. With the constant maddening sway and with leaden heart we teeter precariously on the endless brine and I reflect on the frightening question, how many hundred leagues are we from my beloved dry land? [From an account thought to have been written between 1820 and 1830 by an anonymous passenger, probably aboard an English frigate in the southeast Pacific sailing northwestward.][1]

The high seas are the great stretches of open ocean, deep sea, and seabed lying beyond the exclusive economic zones of coastal states. They make up almost two-thirds of the world's oceans and cover nearly half of Earth's total surface area (about 45 percent). Our unknown traveler conveys something of their otherworldliness and of how unpredictable and terrifying wide open seas can be. Countless ships have been lost to the ocean over the centuries, and with them hundreds of thousands of lives. During World War II, while serving in the merchant navy, my father very nearly became one of those lost lives — on three occasions. On the first voyage he managed to survive his ship's being sunk by a German air raid in the Mediterranean. He and a few other crew members, adrift in a lifeboat, were eventually picked up by an American destroyer. Apart from being cold, hungry, and shaken, they were unscathed. Some months later he survived a second ship's sinking in near-freezing waters in the North Sea. Both escapes were especially fortunate because, being the radio officer, my father had to be among the last to abandon ship. His third brush with a watery death happened when he was boarding a ship due to sail from Scotland to the Pacific. Showing early symptoms of measles, he was ordered to stay ashore. A month later that ship was also sunk by an enemy attack, with all hands lost.

Today the high seas are unowned territory, but if Spain and Portugal had had their way four hundred years ago they would have been claimed, carved up, and possessed by colonizing nations, like great tracts of land were in America, Australasia, Africa, and Asia. This proposition was known as mare clausum (closed sea). In 1609 Dutchman Hugo Grotius set out a radical counterproposition in his seminal pamphlet *Mare Liberum* (free sea, or freedom of the seas).[2] He regarded mare clausum as a kind of ultracolonialism — a grand plan of domination that was immoral, impractical, and contrary to natural law. Grotius believed natural law was "higher law" in which legal principles are rooted in reason, independent of any man-made construct of authority, whether political or religious. He argued that oceans are international territory and that all people have the right to travel and trade across them. Like the air, he wrote, "the sea is common to all, because it is so limitless that it cannot become a possession of any one, and because it is adapted for the use of all." Grotius also hoped to promote a culture of fair-mindedness between nations, believing many potential conflicts could be resolved if countries agreed on codes of behavior to avoid unnecessary war and keep the peace. For this reasoning he is considered one of the founders of international law and admired for his enlightened internationalist outlook: that humankind is better off as a global community, than divided into self-interested nation-states. Like most great ideas, the free sea principle arose from what came before. Grotius drew on elements of ancient Roman law and was influenced by the works of the Italian philosopher Thomas Aquinas, the Spanish theologian Francisco de Vitoria, and the Italian lawyer Alberico Gentili, who was teaching at Oxford University when Grotius was a boy. Thankfully, Grotius's vision largely prevailed over the "closed sea" path. In the twenty-first century we have ended up with a combination of both (although the free sea principle covers the greater proportion of the sea). Today, Grotius's wish for the sea to be free and accessible to all is incorporated as a basic principle of the United Nations Law of the Sea. His "free sea" is now our high seas.

The high seas may be unpredictable, and they can be terrifying, but that doesn't stop us from going there to plunder, pollute, and kill on a catastrophic scale. Every year an estimated 300,000 whales, dolphins, and porpoises die entangled in fishing nets, along with thousands of endangered sea turtles.[3] A longline fishing line carries thousands of baited hooks, and all types of marine life try to eat the bait before it sinks below the surface. More than 300,000 seabirds are drowned by longlines each year,[4] including rare species of albatross, plus turtles, seals, dolphins, and a wide range of other "nontargeted" life. "Fifteen out of twenty-two species of albatross are threatened with extinction. The main threat to albatrosses is death on a hook at the end of a fishing longline."[5]

We are indeed the ocean's grimmest reapers. Across the high seas and beneath them there is more traffic, more mining planned, more underwater noise, and more commercial fishing happening than ever before. It is mega-exploitation of this immense hoard of natural wealth. But it's our right to take it, isn't it? After all, it's part of the global commons.

In theory the global commons comprises shared natural resources that are usable by anyone and cannot be owned. The atmosphere is a commons resource, for example. The air we breathe couldn't belong to a corporation, a country, or an individual (at least we hope not). The high seas and their bounty are a commons resource: the "common heritage of humankind."[6] They belong to no one, yet in a sense they belong to everyone. This is the paradox that has led to the environmental crisis facing the wilderness of plenty.

Raiding Commons Resources

With a "shared" natural resource, let's say fish in the high seas, what actually happens is anything but sharing. In such open access areas, without adequate regulation, individuals, corporations, or countries generally act in their own interest. They exploit wild fisheries to the maximum while stocks are still there. When coastal fishing grounds are exhausted, fishers go farther out to the deep ocean, seeking new species to bring to our tables. Commons resources are similarly drained, and in time they may be wiped out altogether. This means the exploiter gets all the benefits of the irresponsible behavior, while the loss of that resource is borne by the rest of us in the wider global community. The expansion of industrial fishing on the high seas by a handful of wealthier countries is a good example of this problem. As an article in the *Economist* noted, "These wildernesses were once a haven for migratory species. But the arrival of better trawlers and whizzier mapping capabilities over the past six decades has ushered in a fishing free-for-all."[7]

Another tragedy is shark finning, a practice driven by ignorance, poverty, and greed. Apart from the appalling cruelty it inflicts, removing so many hunters from the sea seriously disrupts the predator-prey balance. For example, populations of Humboldt squid have increased dramatically along the Pacific coastlines of Mexico, the United States, and Canada because their predators, such as sharks and tuna, have been overfished. These squid are large and aggressive creatures with voracious appetites that devour great quantities of fish, both commercial and noncommercial species. Communities on the coast of the Sea of Cortez, in the Gulf of California have suffered because of the loss of income from fishing. Harmful fishing methods in these waters aren't new. In 1940 the novelist John Steinbeck and his companion, marine biologist Ed Ricketts, chartered a small fishing boat and sailed along the American Pacific coast to Mexico and into the Gulf of California. In 1951 Steinbeck published an account of the trip, *The Log from the Sea of Cortez*, in which he describes a fleet of Japanese trawlers "dredging with overlapping scoops, bringing up tons of shrimps, rapidly destroying the species so that it may never come back, and with the species destroying the ecological balance of the whole region."

An uncertain number of people in every society are inclined to be greedy and reckless unless they are compelled to be otherwise. Combining this with a compulsion to exploit the undefended natural world for profit produces a hideous mix that is devastating our common ocean heritage. How can we stop it and set in place an alternative normality in which commons resources are used equitably and sustainably? On an institutional and global scale, good practices and enlightened attitudes toward the natural world are fostered through education, agreed-on codes of conduct, incentives for compliance, and passing and enforcing international laws. The UN Law of the Sea, and its subsequent sister treaty, the UN Fish Stocks Agreement, were designed to prevent the high seas from becoming a lawless Wild West; yet, over large areas, that is just what they are—exposed to unregulated or illegal fishing and polluted by both sea-based and land-based industries. Without a cohesive governing structure to manage activities in the high seas we have the silo management approach, with disparate bodies regulating different industries through a mixed bag of international rules, some of them conflicting.

The presumption of good faith—that regulatory bodies would properly implement laws and that sea users would abide by them—was naive and overly optimistic. As Andrew Jillions writes, "Regulation aside, this is still an essentially lawless domain governed by private rather than public mechanisms enforcement."[8]

Regional Fisheries Management Organizations (RFMOs) manage commercial fishing in the high seas and are meant to ensure that fish stocks are plentiful. Currently there are eighteen RFMOs, but the system is fragmented and coverage of the oceans is incomplete. Several of the RFMOs control fishing for one species, such as the Convention on the Conservation of Southern Bluefin Tuna, the Pacific Salmon Commission, and the International Pacific Halibut Commission. Regions of the ocean that have no RFMO coverage are left especially vulnerable to bad fishing practices. Unregulated trawling in large parts of the Pacific, for example, has disastrous consequences for seabed biodiversity and deep-sea fish stocks. In addition, RFMOs are routinely criticized for their inefficiency and for allowing fish stocks to dwindle, in some cases to near extinction. Typically, their primary motivation is making money, not conservation. Conservationists and scientists are frequently exasperated by RFMO managements' decisions. The annual meeting of the South Pacific Regional Fisheries Management Organization (SPRFMO) in 2019 is a case in point. A deep-sea regulation was adopted to allow the New Zealand trawl fleet to continue bottom trawling on sea mounts and undersea ridges in high seas biodiversity hot spots. Over the years these vessels have dragged up tons of corals and other deep-sea bycatch. And absurdly, despite being fully aware of how destructive the trawling is, the SPRFMO has given them the green light to carry on.[9]

RFMOs can't take all the blame for their shortcomings, though, because to some degree they were set up to fail. Some provisions of the Law of the Sea are contradictory, such as the directive to manage fisheries according to their maximum sustainable yield (a system that largely disregards related and dependent species) while also preserving marine environments and wildlife populations. Unfortunately, the prevailing RFMO mind-set remains an economic one, based on the outworn single-species approach, which makes for an inadequate system of controlling commercial fishing in the largest wild region on Earth.

An organization well known for its record of failed management is the ironically named International Commission for the Conservation of Atlantic Tuna (ICCAT). Along with tuna, ICCAT is responsible for conserving related species such as swordfish, marlins, and nine types of sharks. Of the various tuna species found in the Atlantic Ocean, the largest, most highly prized, and most intensively fished is the Atlantic bluefin tuna. They are the superfish—glorious, turbocharged machines built for speed and endurance, able to cut through the water like living torpedoes at almost fifty miles an hour. Nature was lavish with these fish. They are big and they are bold. Atlantic bluefins can grow to twelve feet and weigh up to 1,500 pounds. With a royal blue luster along the back and a pearlescent underside, they are sumptuous and striking creatures. And most unusual for a fish, they are warm-blooded. Unluckily for the bluefins, though, they taste so good that everyone wants to eat them—or at least enough people to have made them an endangered species.

Let's turn to the organization whose very purpose is to protect the Atlantic bluefin and keep stocks at healthy levels: ICCAT. Perversely, to many observers the organization seems bent on wiping out the species. Year after year ICCAT has ignored scientific advice on the catch limits necessary to keep stocks safe, repeatedly setting quotas far above the experts' recommendations. The result is that the Atlantic bluefin population is estimated to have fallen by over 97 percent, and some experts believe the loss is even greater, with less than 2 percent of the naturally occurring unfished population left in the sea.[10] To date, ICCAT's priority appears to lie with keeping the tuna industry happy, not with the well-being of the species. It's a familiar and tedious tale—authorities giving way to the short-term economic demands of the fishing industry rather than ensuring the long-term viability of wild fish populations. Indeed, ICCAT's reputation for mismanagement is so bad that it is known by some as the International Conspiracy to Catch All Tuna.

In 2009 ICCAT finally woke up, if only partially. The Atlantic bluefin situation had reached a crisis. Conservation organizations and some governments were calling for the trade in bluefins to be suspended until the species could rally. At last catch limits were reduced and set according to scientific advice. Bluefin numbers began to pick up, but even as populations made a fragile recovery, ICCAT returned to form by raising the catch quota, despite warnings from scientists and conservationists not to do so. In 2017 the World Wildlife Fund's fisheries project manager, Alessandro Buzzi, expressed his frustration in a statement issued by the organization, which had been urging ICCAT to keep quotas low: "We have been fighting for the last ten years to save bluefin tuna, we are so near recovery that it is a scandal to see ICCAT going back to business as usual; this could jeopardize all the progress we've made."[11]

As part of the global commons, natural resources of the high seas are meant to benefit the whole global community, but there remains a troubling imbalance in the distribution of the wealth they generate. Just five countries—China, Japan, Spain, South Korea, and Taiwan—account for nearly all high seas fishing: 85 percent of it according to Global Fishing Watch.[12] Using these resources in a more responsible and equitable way not only will safeguard high seas wildlife and underwater habitats, but will also improve international relations. How to ensure that has been debated for many years, and it is exasperating to see how long authorities can take to deal with a crisis they have long been aware of. While Rome burns, they say, let's arrange a few committee meetings and consultations. We'll discuss the latest data on collapsing fish stocks and dead coral reefs, spend a few months writing evaluation reports and feasibility studies, then arrange more meetings to negotiate definitions and terms of engagement. After that we'll consider the options, and if we can agree we *might* do something about it.

For several years conservation groups, scientists, and proactive officials have been calling for the Law of the Sea to be brought up to date and into line with greater understanding of natural systems, taking into account new technologies and industries so that the law can better conserve life in the high seas. Law reform is finally in the offing, but it has been a long time coming. The United Nations Division for Ocean Affairs and the Law of the Sea (DOALOS) conducted working-group discussions for nine years before agreeing that more robust law was needed to protect high seas biodiversity. That was before actual treaty negotiations began. Compare that with how swiftly the bureaucratic machinery of war can kick in—within days or even hours. The difference, of course, is how the decision to act is made. Inevitably it will be a prolonged and painstaking business for about 180 countries to reach agreement on such a complex issue. Even allowing for that, however, the urgency of the worsening situation in the global ocean is not getting through. It will take several years to agree on and adopt conservation measures, and snail-paced action planning doesn't suit a rapidly deteriorating environmental situation. The iceberg is looming while the quartermaster casually chats with the first mate about its distance and size when he should be frantically turning the ship's wheel.

Apart from being long-winded and slow, unanimous decision making has other disadvantages. One was highlighted by an attempt to prohibit deep-sea bottom trawling—widely regarded as the most destructive and wasteful method of fishing, since it indiscriminately wrecks underwater habitats, including fish nurseries, spawning grounds, and slow-growing corals. In 2006 United Nations negotiations, supported by a petition signed by 1,100 scientists from around the world, very nearly secured a global ban on the practice. But the ban was scuppered by the veto of one country—Iceland. Of Iceland's sabotaging so crucial an advance in marine conservation Karen Sack from Greenpeace said, "The international community should be outraged that Iceland could almost single-handedly sink deep-sea protection and the food security of future generations."[13]

Once again short-term economic interests overrode the common good. "A nation of 300,000 people stymied the introduction of protection critical to the survival of deep-sea life," writes Callum Roberts in *The Ocean of Life*. "At some point we need to draw the line so that small minorities cannot torpedo the fate of the planet."[14] One way to avoid that is to adjust the rules of collective policy making so that it does not have to be agreed unanimously. Instead a large majority, say 80 or 90 percent in favor should be able to secure a new policy or moratorium. If the system remains as it is, at the mercy of one state that can easily defeat the majority's efforts to save high seas biodiversity, it will be difficult to make much headway.

A New Treaty on the Horizon

The High Seas Alliance is a coalition of thirty-two conservation organizations plus the International Union for the Conservation of Nature, and its aim is to conserve the health and abundance of the high seas. The HSA played a key role in pushing for reform of the Law of the Sea with an additional treaty to address mounting threats to high seas biodiversity. Eventually the contributors' efforts were rewarded. After four days of tough talk at the United Nations headquarters in New York in January 2015, the breakthrough came when governments from around the world agreed on the development, as part of the Law of the Sea, of a legally binding international instrument to conserve the living resources of the high seas (also called biodiversity beyond national jurisdiction, or BBNJ). A few months later the United Nations General Assembly formally set the process in motion by passing resolution 69/292.

The main improvements the BBNJ treaty negotiators plan to make are to ease the creation of marine protected areas and marine reserves in the high seas; to obtain mandatory rigorous environmental impact assessments before allowing a range of potentially damaging activities; and to help developing countries play a greater part in marine science and conservation (known as capacity building). Another important element of the treaty is to bring a fairer distribution between richer and poorer nations of the benefits from the commercialization of substances derived from marine genetic resources such as those used in cosmetic and medicinal products. Leading marine scientist Dan Laffoley and his colleagues sum it up:

> So that we can look at the ocean holistically and in an integrated manner, the Treaty must combine different aspects of ocean management (fisheries management, ocean protection, the managing of mining and shipping). Action is also needed to reform voting rights in sectoral organizations as too many have a few dominant economic stakeholders who can, through consensus requirements, make or break reform and control how decisions are agreed, or which resolutions are adopted.[15]

When so many countries need to reach agreement, progress is slow and compromises have to be made that often result in a weaker legal structure than was hoped for. Despite this, the press reflected tentative optimism in headlines when the new treaty was announced.

UN Gives Green Light for Treaty on High Seas Conservation (*Times of India*, June 2015)

Nations Will Start Talks to Protect Fish of the High Seas (*New York Times*, August 2017)

When I delved into the background of how the BBNJ treaty came to be on the United Nations' table, I admired the negotiators' determination and skill. There are contentious matters to find agreement on between a broad range of countries, each with its own concerns. One example is the question of sharing benefits from marine genetic resources and deciding which governing principle they come under: the common heritage of mankind or the freedom of the high seas. Another is which governance framework the new treaty should have and how it will work alongside existing organizations like the International Seabed Authority. Over two years of preparatory meetings, the draft elements of the agreement have been thrashed out between delegates from the participating countries, related charities and NGOs, lawyers, and scientists. If it is successful, the BBNJ treaty will also enable the international community to reach the Convention on Biodiversity's targets on high seas biodiversity that were set in 2010,[16] plus those in the UN Sustainable Development Goal "Life Below Water."[17]

One morning I had a call inviting me to give an informal talk at a seminar, about the concept of protecting all of the sea and how it could be done. Some weeks earlier I had thrown out the idea to Stephen Eades and David Levy, the coordinator and chair of Marinet, and asked for their thoughts. They were enthusiastic and keen to promote it. To begin with, a summary of the proposal was put on Marinet's website, titled "The Principle of Worldwide Marine Protection: A Paradigm Shift in Our View of the Sea." Beneath the title I added an image of a world map with small red dots in the blue depicting protected areas of the sea to emphasize how small a proportion of the global ocean was safe. Later I went on to develop the proposal further in a pamphlet, *Conserving the Great Blue*, which Marinet disseminated to other conservation organizations to trigger debate and gather support. The idea was seeping out via the website and the pamphlet, and some people wanted to learn more. It was regarded either as an outlandish impossibility or as the obvious way to go.

This invitation wasn't the first time I'd been asked to speak at an event, but as usual I was hesitant. The thought of speaking in public brings on a cold sweat, yet there's nothing more persuasive than hearing people talk enthusiastically about what they hope to achieve and why. I wondered how I could break through this confidence barrier: Take a course perhaps, or practice on friends and family? One evening my friend Clare presented that opportunity. She'd invited a few people for dinner and asked me to join them. I didn't know the other guests, but I decided to go anyway.

As we sat at the dinner table, out of the blue Clare announced that I had a new theory to "share" on how to protect the oceans. My heart sank. "Why on earth did she have to do that?" I thought. People went quiet and looked at me probingly. "Go on, tell them all about it," she said eagerly. I was cornered. Reluctantly, I briefly described the idea. One of her guests, a long-standing environmental campaigner, surprised us with his response. "That's ridiculous," he said abruptly. "You can't stop people fishing!"

"I'm not suggesting people stop fishing," I replied, "I'm saying they should fish in a different way, in a better way so that . . ." I tried to explain, but he wasn't listening and interrupted me in midsentence. "Anyway, how could you possibly control what people do over millions of square miles of remote ocean? It just can't be done," he stated confidently. "But it can be done," I said. "If we can rid the world of smallpox, if we can put people on the moon, if we can set up the United Nations, then we can stop a bunch of ne'er-do-wells from raiding and wrecking the oceans."

"No, no. You can't protect all the sea. All of the sea? That's ridiculous," he repeated, shaking his head and not prepared to hear me out. "It's all protected by international law, so why not?" I asked. "All protected by law!? What law?" he scoffed and continued slowly and emphatically, as if he were talking to a two-year-old. "My dear, I don't know much about it, but I can tell you this," and pausing to take a deep breath, he declared, "you are wrong."

There was a short but screaming silence. I was kicking myself for being ineffectual in the face of this pompous nonsense. How can he say I'm wrong if he doesn't know anything about it? The others didn't know what to say, and for a few moments the room was awkwardly quiet. Then one of them piped up, "Well, I think it's a brilliant idea."

5: Theory to Reality

The classic structure of a persuasive argument for radical change has four main elements: what needs changing, why it needs changing, what we would put in its place, and what we will gain from the change. ACTION I want to add a fifth: how the proposed change can become a reality.

Millions of people around the world understand the first two items in the list—that the sea is in trouble and that something should be done about it. For some, knowledge comes in rather distant forms, through news reports, television documentaries, and newspaper articles. For others, living in coastal communities and dependent on fishing for their food and incomes, awareness of the problem has far more punch, since it directly affects their daily lives. It comes through the personal experience of seeing fish stocks falling, sea levels rising, and waters polluted; of more extreme weather patterns, plus beaches and bays strewn with waste; and of lower incomes and less fish on their plates. The third element is the core of the proposal I make here: turning around the dominant approach to using the sea and creating what some would call a new normal—from maximum exploitation at almost any cost to moderation and respect. Successful regimes can be copied and adopted on local, regional, and national levels. International law could be implemented and enforced as it should have been for nearly thirty years, compelling all sea-using industries to be responsible and accountable. As far as possible, the entire marine environment would then be safe from destructive activities. With this reality, the fourth element of the argument is reflecting on what we would gain. Telling a roomful of skeptics about the advantages could be the way to win them over.

When I unravel the reasoning and anticipate this ideal outlook for the world's oceans, all the advantages lie spread before me like parts of a dismantled clock. There are obvious gains for nature: diminished pollution and cleaner seas, coral reefs and mangrove forests safeguarded, undersea habitats recovering, fish stocks rebuilding, and wildlife returning. They are the same gains that have been documented in hundreds of marine protected areas, but on a global scale. Less apparent is the sea's potential to mitigate the effects of climate change. Robust and resilient oceans can absorb more heat and CO_2 emissions, and they are more able to confront the effects of climate change, such as ocean acidification, than if they are weakened and degraded.

An additional projected advantage is making more of that old favorite—money. The evidence shows, paradoxically, that when marine resources are used well and driven by the law of nature, not by the law of the dollar, there are considerable economic returns for individuals, for businesses, communities, and governments—as the fisheries of Norway and the United States show. Profits generated from well-managed fisheries lift household incomes, and the extra revenue from taxes and exports fills state coffers. The ecosystem approach is key, not only as a management strategy but, perhaps more important, because of the type of thinking it fosters; a mind-set of participation, of collective responsibility, of cooperation and shared advantage. It is a code of rational behavior on how humans can prosper by working with nature, not against it. Following from the financial gains come even greater social benefits. With a perpetually regenerating natural resource such as fish in the sea, those who fish for a living (or who work in related industries like processing and retail) have better food security, longer-term employment, and more to spend on housing, education, and health care. This makes millions of people not only better off and better fed, but healthier and happier too.

The fifth element of the argument for change is trickier: How could all our seas and oceans be protected? I realize there are two elements to the "how" part of the argument. The first is an operational one—how this strategy could be put into practice—and the second is how and why the international community would agree to take that path.

Protecting It All in Practice

Condensed into three stages, the strategy is this: modernize the law, implement the law, and enforce the law. In the words of American antislavery campaigner Henry Ward Beecher, "Laws and institutions, like clocks, must occasionally be cleaned, wound up, and set to true time." As I discussed earlier, the Law of the Sea is out of date. It needs cleaning, winding up, and setting to true time just as Beecher suggests. To a large extent the BBNJ treaty, if agreed to, will do that and will be a great step forward in safeguarding high seas habitats and wildlife. Recognizing the challenges in reaching agreements of this kind and appreciating the considerable efforts put into the process, the treaty's provisions would have to go a good deal further in order to legislate unequivocally for the core of the proposal made here and set our sights on 100 percent ocean protection. Of course, that objective has never been on the negotiating table, and in today's political climate it isn't possible for a new oceans treaty to do more for the sea at this time. I see the BBNJ treaty as a stepping-stone, taking us much closer to where we hope to be in the future. There's no harm, then, in outlining how the Law of the Sea needs amending in order to take the last steps.

The Law of the Sea: Key Reforms Needed

The Law of the Sea's broad-brush provisions, such as "States have the obligation to protect and preserve the marine environment,"[1] presuppose that all the world's seas and oceans are already meant to be shielded from the excesses of human behavior, notably from pollution, damaging practices, and overexploitation. It doesn't say "States have the obligation to protect and preserve *some* of the marine environment," or "States shall cooperate on a global basis in formulating rules, standards, and recommended practices for the protection and preservation of *parts of* the marine environment."

In a modernized Law of the Sea, the underlying principle of protecting and preserving the sea should be reasserted, reinforced, and made plain in terms that cannot be disregarded or misinterpreted: that the health of the global ocean and its life-supporting natural systems should take precedence over commercial interests — everywhere. The reformed law should call for the overhaul of the various administrative bodies and Regional Fisheries Management Organizations into a cohesive governance mechanism to oversee a regime of responsible practice and bring about wider compliance with the law. Standards and rules need to be clearly defined: for instance, What are the marine pollution standards? What degree of fishing damage to an undersea habitat is acceptable, if any? In addition, an updated Law of the Sea should establish a high seas enforcement agency to provide coordinated monitoring and surveillance around the world, with a mandate to board and inspect vessels and withdraw licenses when the law is breached. It could be delivered by an extension of the powers and jurisdiction of the International Criminal Court set up in 1998 by the Rome Statute. Destructive practices would become criminal acts, and the International Tribunal for the Law of the Sea could be given greater powers to take robust action against governments and corporations, and against the individuals who are responsible for ocean-damaging policies and operations (such as allowing land-based industries to pollute the sea or setting fishing quotas above scientific advice).

Some specific articles of the existing law could be enhanced, and others that are no longer relevant could be removed altogether. Article 136, for instance, could be amended from the "Area and its resources are the Common Heritage of Mankind" to include not only the seabed but the water above it — in other words all of the high seas, as Arvid Pardo originally proposed in 1967. Article 116, "Right to Fish on the High Seas," should be revoked to help combat illegal and unsustainable overfishing. Only ships registered with (and regulated by) a regional body would then have the right to fish on the high seas. The anachronistic article 62 (2), which requires coastal commercial fish stocks to be fished to their maximum capacity, should also be revoked (not least because the catch quotas so often exceed scientific advice): "Where the coastal State does not have the capacity to har vest the entire allowable catch, it shall, through agreements or other arrangements . . . give other States access to the surplus of the allowable catch."

Implementing a Modernized Law of the Sea

I'd dwelled long on the complexities of ocean decline and tried to make sense of them. Why are marine environments so misused despite all the legislation to protect them? The usual suspects kept cropping up: weak governance, flimsy implementation, and minimal compliance. They're like three misfits who are always together at any social gathering. When you get weak governance, the other two are always trailing along behind. Laws can't do much on their own: it takes an administrative body to put them into effect, and good governance is essential. If a law isn't properly imposed, how do you know what you're supposed to be doing (or *not* doing) to comply with it? If there are no repercussions when individuals, corporations, or governments break the law, if there is no counter action or penalty, the tendency is to not bother observing it. A modernized Law of the Sea would need a carefully planned and forward-looking governing system to implement a strategy of 100 percent ocean protection. What form would it take?

Regarding oceans and conservation, not only do different bodies manage different sectors in the same places (and not always very well), there are governance gaps too, and where an area is not well covered only partial application of the law is common. Poor governance or no governance is at the heart of the problem, especially in the high seas. Good governance could turn this around, but it's difficult to pin down precisely what that is and how to attain it, especially when it requires an international reach. Getting governance right on land is hard enough. Several countries are quasi-plutocracies disguised as democracies, where policies are as likely to be determined by a cluster of the wealthy and influential as by the will of the general public. The distinguished economist Paul Krugman was interviewed in January 2020 on BBC Radio about his book, *Arguing with Zombies:*

Economics, Politics, and the Fight for a Better Future. He talked about how a blinkered worldview on the part of those in power can block the way to positive change; as an example, he pointed out that "sixty percent of Republicans in the US Congress deny that climate change is happening, and the remainder almost entirely oppose any significant action to limit it."[2]

A joint study by Martin Gilens from Princeton University and Benjamin Page from Northwestern University, based on the analysis of twenty years' worth of federal policy decisions, suggests that the United States is no longer a true democracy and is all but governed by a small economic elite (code for an oligarchy): "Economic elites and organized groups representing business interests have substantial independent impacts on U.S. government policy, while average citizens and mass-based interest groups have little or no independent influence."[3]

In an oligarchy, power lies with a small group of people. In a plutocracy, power lies with the very wealthy, and in many aspects the two systems are indistinguishable. Big business and industry have long had enormous political influence in America, using their multimillion-dollar leverage to sway policy making in their favor—arguably making corruption legitimate and lawful.[4] Increasingly, very wealthy individuals are entering the political sphere in person, and more of the superrich in government is definitely not good for defending nature. Apart from a few exceptions such as Melinda and Bill Gates, the political motives of the extremely wealthy have proved to be self-serving and, in the opinion of many people, amoral. The classic story of billionaires in politics is characterized by their trying to block attempts to redress the vast equality imbalance in American society. Typically they stand against tax increases for the very wealthy, they oppose an affordable health and welfare system for all, and—more pertinent to the oceans' well-being—they repeatedly obstruct proposed environmental protections (such as controlling industrial pollution), they fund campaigns to disseminate disinformation about climate change, and have frequently hindered the state's efforts to reduce fossil fuel extraction. As Luke Darby writes,

If popular support actually influenced public policy, there would have been more decisive action from the U.S. government years ago. But the fossil-fuel industry's interests are too well-insulated by the mountains of cash that have been converted into lobbyists, industry-shilling Republicans and Democrats, and misinformation. To them, the rest of the world is just kindling.[5]

Perhaps the love of money isn't the root of all evil but a kind of sickness that makes people lose their reason. How could billionaires with regressive and pernicious agendas be kept out of (supposedly) democratic government? One way would be to not have billionaires in the first place.

The task is to find the style of governance best suited to regulating our use of ocean resources and to making the laws protecting marine environments work. Global ocean waters are divided into two realms: those within coastal states' two-hundred-mile exclusive economic zones and those beyond, in the high seas. Correspondingly, two areas of law implementation are required; one run by each coastal state as part of its domestic policy (which parties to the Law of the Sea are already committed to do) and the other to cover activity in the high seas. Therefore there is often a double hurdle of faulty governance to jump. First, few countries are fulfilling their UN Law of the Sea conservation obligations within their own waters, and second, as I described earlier, there is no suitable governing structure to implement the law over the rest of the ocean, in the high seas. And this presents another conundrum.

Remodeling the Old or Starting Anew?

The most celebrated exponent of how best to govern commons areas and manage the use of their natural resources is the Nobel Prize–winning American political economist Elinor Ostrom. Her influential book *Governing the Commons: The Evolution of Institutions for Collective Action* is essentially an optimistic antidote to Hardin's pessimistic theory of the tragedy of the commons (which results in an unrestrained free-for-all of excessive use like overfishing and overgrazing). It is a valuable guide for planning strategies on controlling what people do in the sea, within a country's exclusive economic zone and possibly in the high seas too. Ostrom studied communities from all parts of the world that share commons resources such as forests and grazing land, freshwater supplies, and marine fish. She found that people are quite capable of working together to find ways to use these unowned "commons pool" resources equitably and without exhausting them—if they choose to do that (rather than taking many past and present societies' more trodden path of overuse without a view to the future). Over time they agree on rules and codes of behavior for using these resources, some of which have worked well for centuries. Significantly, these rules aren't imposed from on high by a distant and centralized authority; they originate from those who have firsthand knowledge of the resource in question and whose livelihoods depend on its continuing. The spread of community-led marine management across the South Pacific and beyond (as I describe in chapter 8) is a case of practice matching theory. Ostrom identified eight principles for good commons governance:

1. Define clear parameters about who can use the resource and what its physical boundaries are.
2. Make rules adaptable and relevant to local needs and conditions.
3. Allow those affected to be involved in modifying the rules.
4. Make sure the rule-making rights of the community are respected by outside authorities.
5. Agree on a coordinated way community members can monitor the system.
6. Use graduated penalties for rule violators.
7. Provide accessible, low-cost means for resolving disputes.
8. Build nested tiers of governance, from the lowest level up to the entire interconnected system.[6]

Could a commons governance model, which was originally based on using forest resources or grazing livestock, be scaled up to protect the world's oceans? On paper the answer is yes. A structure of trans-ocean governance based on Ostrom's principles indicates that a common-purpose, cross-sector, tiered (but nonhierarchical) system would work best—rooted in the ecosystem-based and precautionary approaches and coordinated by a central agency that could also provide financial, scientific, and technical support for local and regional bodies.

Even if creating the perfect method may elude us, it is possible to find a much improved way of controlling human activities in the high seas that is preferable to what we have now: a jumble of often narrowly focused agencies and with life-depleted, polluted seas to show for it. With several disparate bodies functioning independently and without enough punitive powers, the Law of the Sea's high seas governance system is less than the sum of its parts and is unable to protect the sea and its wildlife. The challenge is to build a cross-sector, coordinated management structure that can.

There have been attempts at addressing the governance problem for some time. Recommendations were made in "Agenda 21: Programme of Action for Sustainable Development" at the Earth Summit held in Rio in 1992, and the International Union for Conservation of Nature published "Ten Principles for High Seas Governance" in 2008. However, despite the diplomatic efforts and good intentions put into such initiatives, little changed. More recently, the Global Ocean Commission has taken greater steps to spur the global community to action. This group of political figures, business leaders, and development experts from all parts of the world came together to work out an all-encompassing strategy of rescue and recovery for the world's oceans, particularly the high seas. Its excellent report "Rescue Package for the Global Ocean," published in 2016, calls for all sectors involved to unify and act to end the cycle of decline. The report describes eight clear proposals for action, the first of which came about in the following year when UN member states agreed that the global ocean was to have a stand-alone Sustainable Development Goal titled "Life Below Water." Targets set include getting more countries to adopt the Port States Measures Agreement to Prevent, Deter and Eliminate Illegal, Unreported and Unregulated Fishing (PSMA; in which authorities can deny foreign vessels access to the port to land illegal catches, refuel, or repair vessels); ending destructive fishing practices; cutting the global fishing fleet back to match available fish stocks; creating a global register of all vessels; banning fisheries subsidies that maintain excess capacity and encourage overfishing; delivering better implementation of protective international law; and making regional fisheries management organizations much more efficient—echoing what conservationists have been demanding for years.[7]

But are the RFMOs' old habits too deep-seated? Their largely nontransparent, single-species, and profit-driven approach can be very persistent and resistant to reform. Rather than trying to graft new thinking onto an outdated construct, it could be better to start fresh with a regionalized management structure that regulates all aspects of ocean use. The commission recommends forming Regional Ocean Management Organizations (ROMOs) to take the place of the regional fisheries organizations. They would manage all activities in a given region, for example, harvesting biological resources (fish stocks, aquaculture, and genetic resources), generating energy (wind power, wave power, oil exploration, etc.), and tackling pollution. However, the commission recognized that it will take many years to see such ROMOs in operation and advises that reforming RFMOs should continue in the meantime.

The ROMO idea is a good one, but a new system of managing what we do in the high seas wouldn't necessarily have to start from scratch. The Regional Seas Programme, for instance, is an international project initiated by the UN Environment Programme (UNEP) to promote cooperation between neighboring countries in conserving marine habitats and living resources. With 143 countries participating in eighteen Regional Seas Conventions and Action Plans, the program spans large areas of the world's oceans (principally in coastal zones). Would it be possible to build a high seas governance structure in the form of a network of such regional bodies that is rooted in Ostrom's principles? Visionary ocean conservation campaigner Elisabeth Mann Borgese described something similar in *The Oceanic Circle: Governing the Seas as a Global Resource*,[8] in which she proposes an integrated style of governance that begins at the coastal commu-

nity level and has support at a national level. Among her other recommendations, one is to establish regional ocean assemblies where
neighboring governments, conservation bodies, civil society groups,
and other parties come together every two years to exchange information, agree on conservation plans, and evaluate progress.

The Anti-Ostrom Approach

The European Union's fisheries management regime, known as the
Common Fisheries Policy (CFP) has routinely breached both its own
primary marine protection legislation (the Marine Strategy Framework Directive) and international law by repeatedly setting catch
limits above the scientific advice. There is no punitive action against
those breaking the law, whether ministers and officials setting exag-
gerated catch limits or vessel owners taking the fish. The regime is
well known for enabling the serious reduction or collapse of many
commercial fish stocks despite laws' being in place to keep them in
good shape. The consequences are dire for marine environments and
for fishing, including for the small-scale inshore fisheries of Europe
and of West Africa, where ships go after they've fished out much of
their home waters. It's absurd that European citizens have been paying public servants to break international law and European law—and
drain the life of their seas. Since a measure of reform was achieved in
2014, the situation has marginally improved, but ministers are still
setting catch limits above scientific advice.

Not to undervalue the sustained efforts of many people to improve the regime, the European Union's Common Fisheries Policy is
a useful guide on a style of governance to avoid at all costs. It sets an
example of how *not* to manage a commons pool resource. Europe's
fisheries were already run down before the CFP's reign, and the tradition of mismanagement has carried on for decades. This is a story of
how government policies have allowed numerous fish populations to
plummet, many marine habitats to be weakened or demolished, and
small fishing communities to fall into decline. Could analyzing failure
point the way to success? Here are some characteristics of Europe's

flawed formula for managing marine commons pool resources. Identifying the opposite of each could reveal the elements of an alternative system of governance that works.

Prioritizing Profit

Prioritizing profit is characteristic of governance that puts corporate profits before preserving the living resources that generate them. Evidently ministers and officials have been representing the fishing industry and not the taxpayer, the marine environment, or its wildlife. In *The End of the Line*, Charles Clover writes, "The only political influence over the sea's bounty is exerted by an industry—the wild capture fishing industry which in Britain is roughly the size of the lawnmower industry. No one would dream of allowing the lawnmower industry to dictate the policy of a sovereign state or even a federation."[9] Taking the opposite approach gives us the first element of good governance. It is to underpin all policy and management planning with the precautionary and ecosystem approach, giving the health of marine environments and resilience of wild populations priority over vested economic interests. Therefore, knowing the industry, and the lobbying and reelection pressures on them, politicians should be taken out of fisheries management.

Having Sketchy Knowledge, yet Disregarding the Science

When setting catch limits for each fish species, ministers and bureaucrats have often demonstrated a bewildering level of what might be labeled arrogant ignorance by paying little heed to those who know how many fish must be left in the sea to ensure a stock's long-term viability. Instead, industry's demands frequently override scientific advice. Many catch limits are set beyond sustainable capacity, meaning fish populations are not able to breed sufficiently to maintain the stock. To understand this you don't need to have a degree in biology or be an experienced fisher, but you do need a little bit of sense. Callum Roberts notes that "it is a system of competitive bargaining in which every fishery minister competes to get the best deal possible for his or her country's fishing industry."[10]

Targeting single species for catch limits reveals that the decision makers lack understanding of undersea biology and natural processes. In a mixed fishery myriad forms of marine life mingle in the water—fish, mammals, birds, reptiles, invertebrates. Consequently, many types of sea life are caught and killed in the same nets as the target species, making a genuine discard ban nearly impossible (accidental bycatch happens less often with pelagic schooling species like herring). Also, because of complex predator-prey interconnections between marine animals, large catches of one species can be significantly detrimental to another. All the conservationists and scientists I've spoken to believe that politicians should not be setting the rules on matters they know little about, not least because their interest is generally short-term. The second element, therefore, is don't ignore the science—be guided by it.

High Command, Ivory Tower Rule without Transparency

The Common Fisheries Policy is run by a top-down, centralized body with little knowledge of local concerns and regional differences. Meetings of the fisheries ministers are closed to the public. Ministers are given advice by their own ministries, who in turn have been told what the fishing industry wants. It is a heavily bureaucratic process of decisions made behind closed doors, short on accountability to those who fund policies and pay the salaries of policy makers: taxpaying citizens of the European Union. The contrary management style includes representation from all interested sectors and stakeholders in decision making.

Illegality and Misappropriation

Common Fisheries Policy practice is at odds with the European Union's overarching law on the marine environment, the Marine Strategy Framework Directive, and as a result the European Union breaks its own laws meant to protect seas and marine life. And since December 2019, ministers are even breaking Common Fisheries Policy law, as they continued to set catch limits above scientific advice despite the 2020 deadline to keep all fishery quotas at sustainable levels. Law administrators—politicians, civil servants, or whatever you call "managers" of the state—are responsible for the wise management of natural resources on behalf of its citizens. But is the Common Fisheries Policy unjust privatization of an unowned commodity? Most of the catch quota is allocated to a handful of large companies and leaves owners of small inshore fishing boats with slim pickings. Almost one-third of the quota is divided between companies owned by families on the UK's *Sunday Times* "Rich List." "Some of these families have investments in dozens of other fishing companies, meaning companies holding thirty-seven percent of UK quota are wholly or partly owned by these Rich List families."[11] This is the unlawful appropriation and unfair distribution by stealth of a commons resource. Wild fish in the sea do not belong to those governing the state, nor do they belong to wealthy people in the fishing industry. The opposite of this is to ensure that policies adhere to the law and that resources are managed for the common good, not for a minority of vested interests. (The United Kingdom has left the EU, and CFP regulations no longer apply. Will British fisheries be managed any better, we wonder?)

An Inability to Adapt

With the Common Fisheries Policy there is resistance to adaptation or reform. There have been some improvements since the reform of 2014, but long-established routines are hard to shift, and in spite of pressure from conservation organizations, scientists, the media, and other concerned parties, ministers continue to ignore scientific advice on many stocks.[12] The most effective environmental management systems, whatever the scale, are flexible and responsive to changing conditions and a greater understanding of natural processes. In this way the system will always be improving. They also adhere to regulations designed to stop overfishing.

Unsurprisingly, the "anti-characteristics" of the Common Fisheries

Policy by and large match the key features of good governance for using commons pool resources recommended by Ostrom, Mann Borgese, and the Global Ocean Commission. In time we hope to leave failed approaches in the past and enter a more advanced era of commons resource use with the kind of governance that guarantees a more equitable share of benefits, listens to those who understand best, is inclusive and transparent in decision making, and uses the force of the law to keep oceans safe. Let's suppose such a governance system exists, how could it ensure compliance with the law over the vast and wild expanses of the high seas?

Enforcing the Law

Earlier I described the achievements of the Commission for the Conservation of Atlantic Marine Living Resources to give an example of the ecosystem-based approach to marine management in practice and to see whether it works as predicted on paper. In large part CCAMLR also answers the question of how to enforce the law across the high seas, because all the waters under its jurisdiction are in the high seas, encompassing over 13,800,000 square miles of ocean, which is about 10 percent of Earth's total ocean surface. So how do they control what happens over such an enormous stateless area?

I went to the headquarters of the British Antarctic Survey on the outskirts of Cambridge to ask that question of Dr. Mark Belchier. He is science manager of the ecosystems team (providing fisheries advice), and at that time he was the chair of CCAMLR's Scientific Committee. Dr. Belchier greeted me with a smile and a firm handshake, then we walked toward the cafeteria, past walls hung with dramatic photographs of Antarctica: a tiny figure in the snow, dwarfed by walls of pale blue ice; a queue of Adelie penguins leaping into the sea; whitewashed mountain peaks overshadowing the old whaling station on South Georgia Island. In the seas around Antarctica the laws of nature have overthrown the primacy of profit. It's one thing to switch the underlying principle of management in favor of conservation, but it's quite another to get people to adopt new thinking, to alter their behavior, and to follow the rules.

The cafeteria was busy, but we found seats at a table on the far side. My first question for Dr. Belchier was, "How does CCAMLR get the fishing industry to comply with regulations in the Southern Ocean?"

We use various monitoring and compliance controls. Within the CCAMLR region, every fishing vessel is licensed to operate and has to stick to the regulations to carry on fishing. For example, bottom trawling is banned everywhere, each vessel must have an independent observer on board a hundred percent of the time to monitor what goes on when they are fishing, and he or she has to be of different nationality to where the ship is from. All ships have to be fitted with a Vessel Monitoring System (VMS) and an Automatic Identification System (AIS) for surveillance and tracking their positions at sea. There are also a number of patrol vessels from New Zealand, Australia, France, and the UK policing the region.

Dr. Belchier went on to explain that the catch documentation scheme provides a paper trail to prove that fish were caught legally, making it difficult to land and sell an unlawful catch. Illegal, unregulated, and unreported (IUU) fishing hasn't been eliminated, but it has been substantially reduced. Fish catches by IUU vessels in CCAMLR's jurisdiction fell from approximately 44,000 tons a year during the 1990s to under 2,200 tons a year by 2011.[13] The battle against illegal fishing is ongoing. Despite the drop in illegal catches, it remains a major cause for concern. Other management areas that CCAMLR is working to improve on are reducing wasteful fishing and gathering more accurate data.

Besides toothfish and krill and their dependent predator species, there was great concern for the number of seabirds that were dying as accidental bycatch. Birds were being dragged underwater as they tried to eat bait hooked onto longlines dropped from the stern. When fairly simple changes in fishing methods were introduced to reduce seabird deaths, the number of recorded deaths fell from about seven thousand in 1997 to fewer than ten by 2012. Over the past fifteen years, seabird mortality from fishing operations has been reduced from thousands of birds annually to almost zero in fisheries regulated by CCAMLR.[14] The conservation measures include weighting baited hooks so they sink beneath the surface before birds can reach them; placing banners and streamers around the area of the vessel where lines enter the water; not dumping offal left from processing over the side while fishing (only when the vessel is steaming); and setting fishing gear only at night.

Well-enforced rules and regulations are essential, but Dr. Belchier stressed that the best way to compel industry to fish wisely and respectfully is by driving out old approaches and fostering a modern attitude toward taking living resources. Education is crucial. As fishers begin to use improved methods, a different cultural perspective evolves, and more progressive systems become the normal way of operating. I ended by asking whether he thought the CCAMLR model of regulation and enforcement should be expanded right across the high seas, forcing responsible fishing in the whole ocean, beyond national exclusive economic zones. "Yes I do," he said, "all the regional management organizations should use ecosystem-based management and always give priority to the health of the ocean's wild populations and ecosystems."

In *The Ocean of Life*, prominent marine conservationist Callum Roberts concurs. "What we need is governance like Antarctica's to be rolled out across all of the high seas and for it to apply to everyone, whether or not they have signed up to international conventions or treaties."[15]

In a sweet twist, technological advances that contributed so much to the collapse of many commercial fish populations are now all-important in rebuilding them. For more than a century, fish have been under attack from a series of developments in the industry. It began with vessels moving from sail power to steam, followed by onboard refrigeration, mechanized net-hauling gear, and sonar equipment to pinpoint where the fish are. The keenest fish hunters will even fly a plane over the sea to spot schools of high-value fish such as tuna. It's time for a technology counterattack on the side of the fish. Global Fishing Watch (GFW; https://globalfishingwatch.org/) is an international nonprofit organization set up jointly by Google, the nonprofit environmental watchdog SkyTruth, and the marine conservation organization Oceana. Using satellite data GFW creates a near real-time online map to track the locations, identities, engine power, tonnage, crews, voyage records, and more of over 60,000 commercial fishing vessels around the world. In an industry marred by unlawful fishing, human rights abuses, animal cruelty, and corruption, GFW is pushing for more transparency and information sharing to counter ocean-going crime and to advance conservation. It enables any internet user to monitor commercial fishing across the world's oceans, and the data is freely available to all. Seafood suppliers can see where and how fish were caught and check that their products are well sourced. Fishers can show they are fishing responsibly, which adds market value to their catch, and can alert the authorities to other vessels' suspicious activity. Scientists, students, and journalists can analyze the impact of commercial fishing on the marine environment (for example, the success of a new conservation regulation). And most important, with the information that Global Fishing Watch gathers, authorities are able to do more to stop illegal fishing.

Law-enforcement methods include preventing vessels from landing catches in port, withdrawing operating licenses, impounding catches, prosecuting crew members or vessel owners, and imposing fines or custodial sentences. At sea, coast guard and naval vessels can be part of a patrolling force, monitoring activity and intercepting vessels breaching regulations. For instance, GFW can reveal incidents of transshipment, when one vessel offloads its illegal catch to another far out at sea (to avoid port landings), making it difficult to track catches. Global Fishing Watch is also an enormously valuable tool for policing marine reserves and marine protected areas (such as spotting trawling in an area where it is banned).

There are other satellite monitoring and enforcement organizations taking on the illegal fishing trade, such as Fishspektrum and Interpol's Global Fisheries Enforcement team, which supports enforcement agencies in Interpol's 192 member countries. Illegal activity isn't confined to fishing; the vessels involved are frequently used to traffic drugs and people too. Interpol notes, "We work with our member countries along all points of the fisheries supply chain, both on shore and at sea, to raise awareness about the impact fisheries crime can have. Our intelligence officers aim to tackle entire criminal networks, not just individual poachers."[16]

The power and value of initiatives like Global Fishing Watch depend on what the authorities do with the information available to them. Knowing about illegal operators—who they are and where they are—enables regulating bodies to take action and impose penalties harsh enough to stop criminals and discourage other offenders. Whether that happens often enough to stamp out illegal fishing, I suppose time will tell. But certainly the argument that it is impossible to know what ships are up to on the high seas is dead in the water. Governments could be taking greater advantage of these fairy godmother technologies by establishing more monitoring, surveillance, and enforcement programs. Information sharing and cooperation between authorities are also invaluable in the fight to defeat illegal and irresponsible fishing.

In conclusion, on the question "How?" the means to implement and enforce ocean protection law, even in the most remote places, already exists. The next question might be "How much would it cost to protect the entire global ocean?" A new system of governance, enforcement controls, inspectors, patrol boats, satellite tracking, and such all sounds expensive. It's difficult to put a price on establishing and maintaining a marine protected area. There are a great many variables such as location, size, and the level of protection given. In 2004, one study suggested a figure of US$12.4 to 13.9 billion a year to provide 30 percent ocean coverage.[17] Using a combination of sources and allowing for inflation, an aggregated projection puts the annual running cost of protecting 30 percent of the ocean at between US$20 and 23 billion. Scaling that up would give a yearly figure of between US$66 and 77 billion to safeguard the whole global ocean. If you consider that in 2018 global military spending was US$1.8 trillion[18] (almost twenty-six times as much), the cost of keeping the world's oceans safe from harm is a bargain. And most important, it would cost much less to prevent a catastrophe in the making than to try to put things right afterward. Besides, much of that figure would be offset by revenue generated from licensing industries and from more plentiful and profitable fisheries. Added to that, hundreds of thousands of jobs would be created: for scientists, administrators, fishers, law enforcers, technicians, and more.

How Would the International Community Agree to 100 Percent Ocean Protection?

It's hard to imagine a throng of delegates from all (or most) countries of the world sitting around the table at the United Nations and signing a dedicated international agreement that overturns a centuries-old human-centric worldview of the sea to enshrine in law that the world's oceans are primarily places to conserve and secondarily places to exploit—and then only with strictly controlled permissions. If and when such a change does occur, perhaps it will be by a different path, taken step by step and over time. Gradual advances will come with tighter regulations on polluting industries on land and at sea and more enlightened fisheries regimes. A few national fisheries management policies are moving toward a precautionary, ecosystem-based approach within their EEZs, and they are reaping the rewards. History shows that when a better way of doing something emerges in one place, other people, other countries, other societies follow suit. So it's reasonable to suggest that fishery productivity in places like Norway and the United States is likely to trigger copycat improvement in other countries until eventually most human pursuits in the two-hundred-mile band seaward of the world's coastlines will be managed more wisely than they are now. That said, there may not be time to wait for incremental change.

incremental change?

As I understand it, the most pressing risk to the sea and its wildlife—indeed to all of the natural world—is institutional and political inertia. With bold government action, all the rest is solvable. There are welcome signs of greater political engagement, although as a general rule governing authorities are slow to react. They cling to a misguided allegiance to business and industry and all too often are led by individuals with little integrity or vision. This position and perspective pervade government at local, national, and international levels and means that the deterioration of the world's oceans doesn't stop.

Like me, roughly two-thirds of the world's population are fortunate to live in some kind of democracy, where we choose who governs us. We make a kind of deal with these people. With our votes we give them a mandate to manage the affairs of state, and with our taxes we pay them to do it. It's a contract between citizens and state: it's a job. A significant part of the job is to manage the use of natural resources well. That could entail making sure ancient forests are kept standing, rivers and oceans are unpolluted, wilderness areas are left alone. It includes safeguarding nature's life-support systems, which provide us with breathable air, a habitable climate, food security, and a clean water supply. But in scores of cases governments break that contract. It's all too evident in clear-cut forests, polluted cities, and emptied seas. Meanwhile, those in power still take their handsome salaries from the public purse. It's a lose-lose situation for Jane and Joe Public and for Mother Nature.

I wanted to know how governments could be made to wake up and take decisive action to resolve the ocean crisis. And what people could do to see that international laws to protect the sea and its wildlife are properly implemented and enforced.

In 1983 Ansel Adams, legendary landscape photographer and passionate defender of the American wilderness, was interviewed by David Sheff for *Playboy* magazine. One of the things he said was, "It is horrifying that we have to fight our own government to save our environment." He was probably right. But how do we do that?

6: Counteroffensive

In 2013 a group of 886 Dutch citizens represented by the Urgenda Foundation took their government to court for not taking enough action to cut greenhouse gas emissions in order to address climate change. They argued that the state has a duty to protect its citizens from the anticipated catastrophic consequences of excess greenhouse gases being pumped into the atmosphere, especially carbon dioxide. The group's attorney, Roger Cox, explained, "What we are saying is that our government is co-creating a dangerous change in the world. . . . we feel that there's a shared responsibility for any country to do what is necessary in its own boundaries to mitigate greenhouse gas emissions as much as is needed."[1]

The plaintiffs were people of all ages, from all walks of life. On June 24, 2015, one of the three judges read out their final verdict, and the courtroom resounded with jubilant cheers.

> The State must do more to prevent the threats caused by climate change, given its duty to care for the protection and improvement of the environment. Effective control of the Dutch emission levels is a task of the State. . . . The State cannot hide behind the argument that the solution of the global climate problem is not just dependent on Dutch efforts. Every reduction in emissions contributes to the prevention of dangerous climate change. As an industrialized nation, the Netherlands should be a frontrunner in this respect.[2]

People hugged each other, elated with relief. They had won. The judges agreed with the plaintiffs and declared that government plans—to cut the emissions 14 to 17 percent below 1990 levels by 2020—were illegal, and they ordered the Dutch government to initiate actions necessary to cut them by at least 25 percent by the end of 2020. After the verdict, Roger Cox was overcome. He told the press tearfully, "We have worked so hard on this case, and it's of such importance to our society."[3] The judgment was reported around the world, hailed as "a landmark ruling," and it has inspired similar actions. Climate change lawsuits against governments and fossil fuel companies have been initiated elsewhere, including the United States, Norway, Pakistan, Colombia, Indonesia, Sweden, and Belgium. The Dutch government appealed the ruling, but the Court of Appeals upheld the district court's decision. The second appeal was to the Supreme Court, and in December 2019 it also upheld the ruling, making it the first time a supreme court ruled to defend citizens' human rights in the face of government's failure to tackle the causes of climate change.

When people worry enough about their future and their children's future, when they are exasperated enough by state inefficiency or negligence, they become disaffected and will protest, whether in the courtroom, at the ballot box, or on the street—made clear with the many thousands of citizens joining climate change protests around the world, taking part in all kinds of nonviolent civil disobedience events to get the message through. The victory of the Urgenda climate change case is a model of proactive citizenship power that could force a government to comply with its obligations under international law. Treaties are statutory law—legally binding written rules of conduct. There is also customary law, a code of practice that evolves over decades, even over centuries. Treaties apply to the participating countries, whereas customary international law is binding on the whole international community, developing from a consensus on what is regarded as civilized behavior.

An Ancient Precept Comes to the Rescue

The public trust doctrine is a principle of customary law dating back to ancient Roman law. It is embedded in the laws and constitutions of several countries, including Australia, Brazil, Canada, Ecuador, India,

Kenya, Nigeria, Pakistan, the Philippines, South Africa, Sri Lanka, Tanzania, Uganda, and the United States. The doctrine entrusts governments with managing natural resources and the commons in the best interests of their citizens: the ones alive today and those who will live in the future. Frustrated by the authorities' lack of response in dealing with climate change, people are beginning to band together and use this little-known legal principle to challenge their governments in the courts. The doctrine asserts that government should act as a responsible trustee, making sure that the natural capital "fund" of the Earth is respected and maintained in the long term because, as citizens, we are the beneficiaries of the fund and all entirely dependent upon it. It has three simple elements:

1. Common natural resources cannot be privately owned, and instead are held within a public trust.
2. Government authorities are trustees of the trust and therefore must manage it wisely.
3. The beneficiaries of the trust (both present and future citizens) can hold the trustees accountable for the mismanagement of the trust.

The public trust doctrine has been used to protect natural habitats in several court cases, one of the best known being *Illinois Central Railroad v. Illinois* in 1869. When the government of Illinois granted a private railroad company title to Chicago's lakeshore and waters a mile out from its coast, the Supreme Court overturned the transaction, declaring that the lake and the ground beneath were protected by "a title held in trust for the people of the State that they may enjoy the navigation of the waters . . . and have liberty of fishing therein, freed from the obstruction or interference of private parties."

As a protector of nature, the public trust doctrine is emerging as a potential force for major change. In 1970, American law professor Joseph Sax reinvigorated the quiescent principle in his article "The Public Trust Doctrine in Natural Resource Law"[4] (and in his book *Defending the Environment*, published the following year). Sax urged citizens to use the doctrine as a means of pressuring authorities to protect and preserve natural resources in the wider public interest, not for the benefit of a minority. He broadened the doctrine's scope beyond rivers and seashores to include other natural habitats and features: land, air, wildlife, forest, lakes, wilderness, archaeological sites, and oceans.

More recently, echoing Sax's logic, Mary Christina Wood proposes using this ancient principle to highlight governments' environmental responsibilities. Her book *Nature's Trust* is an eloquent call to action, encouraging people to hold governments to account for mismanaging and squandering natural resources: "Those who allow the destruction of natural resources to advance singular and corporate interests, or to serve their own political or bureaucratic interests, act in violation of their duty."[5] Wood argues that the judiciary should order the executive and legislative branches of government to protect natural resources for the common good.

The Philippines provides another auspicious example of the public trust doctrine's winning the day. In 1999, attorney Tony Oposa Jr. filed a suit on behalf of the NGO, Concerned Residents of Manila Bay against eleven Philippine government agencies (including the departments of health, the environment, water, and the police) for polluting the bay. In 2008 the Supreme Court ruled in favor of the residents, finding that the agencies were failing in their statutory obligations to safeguard the bay's waters, and ordered them to clean up, rehabilitate, and preserve Manila Bay.

The combined area of the territorial waters of countries where the doctrine is written into the constitution amounts to almost 28 percent of all exclusive economic zones, meaning that if those countries properly observe their own laws, fifteen million square miles of the global marine environment would already be well managed and safe from harm. Even if the public trust doctrine is not formally incorporated into statutory law or the constitution, it still stands as a basic principle of any government's duty of care, to regulate the use of natural resources in the best interest of all citizens. As the Philippines Supreme Court concluded in the Manila Bay case, "the right to a balanced and healthful ecology need not even be written in the Constitution, for it is assumed . . . to exist from the inception of mankind."

Climate litigation cases using the public trust doctrine are on-going in several countries, among them, Belgium, Colombia, India, and Pakistan. In the United States, the NGO Our Children's Trust has filed climate change lawsuits on behalf of groups of young Americans against governments or corporations in several states. The young-sters claim that promoting the use of fossil fuels and failing to im-pose regulations needed to reduce CO_2 emissions and avert climate change violates their constitutional and human rights to live in a safe environment in the future.

It's yet to be seen if a court ruling in favor of the plaintiffs will alter government energy policies. Lawsuits are expensive and protracted affairs, often taking years to conclude and with no guarantee of even-tual success—which doesn't befit a climate emergency. This is leaving aside the absurdity of citizens' having to take legal proceedings to compel authorities to do what they are legally, constitutionally, and morally bound to do anyway. But if it works, well and good—that's what matters.

Today's high-level politicians, and CEOs and board members of chemical and fossil fuel companies won't have to live with the reper-cussions of poor policy or management decisions they've made in re-cent years or are making now. They're mostly wealthy men in their fifties, sixties, and seventies and will be dead before very long. The younger generation will be dealing with the consequences of these people's incompetence if they have to live in a polluted and unstable world in years to come.

Since I began, I had certainly been learning much about international environmental law and especially about what it should be doing for the sea. So far, customary law is doing little, and even big-gun inter-national agreements like the Law of the Sea and the Convention on Biodiversity have not kept oceans safe. I wondered, then, If these laws weren't protecting nature as they should, was there another one that could?

One morning I had a phone call from Stephen Eades, the coordi-nator of Marinet, suggesting I look at the Montreal Protocol as a possible model for a new oceans law. I was curious and did as he asked. The original 1985 framework was the Vienna Convention for the Protection of the Ozone Layer, and the Montreal Protocol was the implementing treaty, negotiated two years afterward, with twenty-four countries and the European Union coming to agreement in September 1987. Its purpose is to protect Earth's stratospheric ozone layer from the man-made chemical substances that were destroying it. Stephen had watched a documentary film, made for the Public Broadcasting System in the United States and Channel 4 in the United Kingdom, called "Ozone Hole: How We Saved the Planet." It tells the story of how and why the Montreal Protocol was agreed on, the individuals who made it happen, and why it is widely considered the most successful environmental treaty.

The ozone layer is gas in the upper atmosphere that shields Earth from the sun's harmful ultraviolet radiation, which causes skin cancer and cataracts, suppresses the human immune system, and destroys agricultural and natural ecosystems. Without the ozone layer, life on Earth would eventually cease.

At first negotiators and technicians focused on eliminating chlorofluorocarbons, which were used in many products like aerosol sprays and refrigerators. Over the years several amendments have been made to include other ozone-depleting substances including halons, carbon tetrachloride, and hydrochlorofluorocarbons. Since coming into force in 1989, the Montreal Protocol has phased out over 98 percent of these harmful chemicals, and the infamous hole in the ozone over Antarctica is closing up. It is also the first international treaty with full participation and ratification by all the countries of the world, and it has almost full compliance.

The Montreal Protocol works—but why? How is it different from the Law of the Sea? One of the essential features is its flexible and dynamic process, which enables the law to adapt to new knowledge and technologies, meaning that additional ozone-depleting substances have been added to the original list of those to be phased out. The Protocol adopts a precautionary approach, and it is founded on the principle of common but differentiated responsibilities. This is when all parties agree on a common goal but recognize that some countries will need more time and financial assistance to reach targets. To offset the costs of making the transition to ozone-safe alternatives, wealthier countries contribute to a multilateral fund made available to less well-off countries. The Protocol has a Scientific Assessment Panel of independent experts who are not politically affiliated and who are sworn to objectivity. Over the years the panel has generated a culture of solid trust in the science. There is also an Environmental Effects Assessment Panel and a Technology and Economic Assessment Panel, to assist participating countries in the practical aspects of reaching agreed treaty targets.

Another important element is that the law controls and ultimately stops the production of damaging substances instead of trying to prevent them from escaping into the atmosphere, since you don't have to deal with an offending substance if it doesn't exist in the first place. A friend commented, "Yes, it saved the ozone layer, but banning a gas in an aerosol can is quite a simple problem to solve." Not so. Dr. Stephen O. Andersen, senior founding cochair of the Protocol and director of research at the Institute for Governance and Sustainable Development, was assigned the task of finding ways to manufacture products without chlorofluorocarbons. As one of the contributors to the documentary, Dr. Andersen said this: "It was the most miserable job someone could get. It's hard to believe how many things we had to change. There were medical applications, weapons applications, rocket manufacture, foam-blowing agents . . . everything imaginable from a refrigerator to a space capsule: 100 chemicals and 240 sectors!"[6]

I like a success story, most of all in the realm of nature conservation, and the Montreal Protocol provides a good one to deconstruct and learn from. A crucial component was having sound and reliable evidence detailing the growing threat to the ozone layer. This was communicated with one united "voice of science" to convince politicians and policy makers that they had to take action, and fast.

At Stephen Eades's suggestion, I emailed Dr. Andersen and briefly explained that I was researching the "possibility of protecting the entire global ocean from industrial misuse and overexploitation . . . fully recognising the damage we are doing to it, in much the same way

[handwritten margin notes: "part of the problem is convincing people that this is a real issue + action needs to occur fast!"]

we had to do for the atmosphere and specifically, the ozone layer." I asked Dr. Andersen if he would meet me and "discuss the kind of legal mechanism we would need to safeguard all seas and oceans, using the Montreal Protocol as a model." He replied the very same day, agreeing to meet, and I was delighted. The timing of my email was serendipitous because Dr. Andersen and his colleague, Dr. Suely Carvalho, a senior expert member of the Montreal Protocol Technology and Economics Assessment Panel, happened to be visiting England the next month, so I wouldn't need to make the journey to Washington DC, where he's based. I sent him the pamphlet I had written outlining the ocean proposition, and we arranged to meet three weeks later at the hotel where they would be staying, just outside London.

When the day came for me to go to London, I had changed my mind about the purpose of the meeting. I decided not to ask Dr. Andersen if he thought the Montreal Protocol could be used as a model for an oceans treaty—that would take far too long to achieve. Instead, I planned to ask this: "What can the Montreal Protocol itself do to help protect the world's oceans and marine life?" In other words, Can we use what we already have? Especially as it has proved to be so effective.

I met Dr. Andersen and Dr. Carvalho in the garden of their hotel, and early into the meeting I asked that question. It seemed to ignite the conversation. They had read my proposal and talked it over the previous day and were thinking along similar lines, believing there is indeed potential for the Montreal Protocol to do something for the sea. Their attention turned to hydrochlorofluorocarbons and to carbon tetrachloride, which is currently banned as an emission but not as a component in manufacturing (called a feedstock). These chemicals are used in the manufacture of some of the plastics that are blighting oceans and killing huge quantities of marine life.

In the report that Dr. Andersen and his coauthors subsequently produced, he wrote, "The Montreal Protocol can take on the phase-out of ODS [ozone-depleting substance] feedstocks used in manufacturing plastics that are harmful to oceans once the Parties appreciate the opportunity and confirm the availability of alternatives and substitutes to plastics."

I'm very happy about the outcome of that conversation and the motivating influence it had on Dr. Andersen and Dr. Carvalho. As I write this, they are completing for publication a road map on just how the idea of finding synergy in environmental treaties can accomplish more than one goal at once, and do it faster.

Prohibiting a substance takes time. The proposal is presented for debate at the treaty conference and has to be agreed on by all parties in order to become law, so any such ban could not happen overnight. Nevertheless, the possibility of expanding the scope of the treaty to defend the world's oceans is promising and exciting.

Inspired by Dr. Andersen and his colleagues' plastic attack strategy, I wondered if the Montreal Protocol could come to the rescue on other battlegrounds of ocean assault. Two villains of the sea immediately appeared in my mind's eye, like a couple of reprobates in a police lineup. They are oxybenzone (which may also be labeled benzophenone-3) and octinoxate, which are found in most sunscreens currently on the market.

On sunny weekends and summer vacations thousands of us stream to beaches and coastal resorts, slap sunscreen all over ourselves and the kids, lie in the sun for a while, then go for a swim. When creams containing oxybenzone and octinoxate are washed off, they disperse into the sea and harm marine life. (Taking a shower or a bath will also wash the chemicals into the natural water system and into the sea.) Even in very dilute quantities these chemicals (found particularly in spray-on sunscreens) disrupt coral reproduction and growth and exacerbate coral bleaching. Coral reefs have the greatest diversity of species of all Earth's habitats, and they face multiple lines of attack: overfishing and destructive fishing (such as trawling and dynamite fishing); pollution; acidification; and rising sea temperatures, which lead to coral bleaching. This damage happens when coral becomes stressed by temperature or chemical changes in the water. The coral then expels its microscopic algae inhabitants, which it needs to survive. Algae give corals their color, too; when they are gone the coral turns white and will eventually die unless the stress factors abate. In short, if oxybenzone and octinoxate don't kill corals, they will make

[handwritten marginal note: a combination of treaties can be effective]

them sterile, yet an estimated 15,400 tons of sunscreen lotion enters reef areas around the world each year.[7] Millions of marine species depend on coral colonies, which makes coral bleaching a very serious problem. All coral reefs are at risk, and in 2016 and 2017 Australia's Great Barrier Reef was hit with mass bleaching events and about two-thirds of the reef's northern sector died off. It used to be thought that coral bleaching was solely a consequence of climate change, but a study published in 2015 found that oxybenzone from sunscreen contributes to the problem. As Dr. Omri Bronstein of Tel Aviv University explained in a press release, "We found the lowest concentration to see a toxicity effect was sixty-two parts per trillion—equivalent to a drop of water in six and a half Olympic-sized swimming pools."

In the waters of the US Virgin Islands, researchers recorded concentrations of oxybenzone twenty-three times higher than the minimum levels that are toxic to coral. "Current concentrations of oxybenzone in these coral reef areas pose a significant ecological threat," Dr. Bronstein warned, adding, "although the use of sunscreen is recognized as important for protection from the harmful effects of sunlight, there are alternative sunscreens, as well as wearing sun clothing on the beach and in the water."[8]

This is the maddening heart of the matter; we know that coral reefs are home to the ocean's richest biodiversity, yet a chemical known to severely harm or kill marine life, and particularly corals, is widely used in countless products that are readily available for sale around the globe. Why is this possible? People are buying these products, applying them, and unknowingly contaminating the sea and poisoning wildlife every day. However you view it, it's madness. There are a few places where harmful sunscreens are banned already, or soon will be, and the number is growing, but not fast enough. They include Hawaii, the Caribbean island of Bonaire, the Pacific island of Palau, Key West, Florida, and some parts of Mexico.

With these deadly goings-on, out of sight and out of mind, I put it to Dr. Andersen: Could the Montreal Protocol also phase out the production and use of these sunscreen chemicals that are wreaking serious damage in the sea? I took his reply as favorable.

So you may have identified an opportunity for more specific and environmentally appropriate advice going beyond sun protection to exactly *what* sun protection. The communication institutions of the Montreal Protocol are up to this. And the environmental authorities protecting the ozone layer could certainly cooperate with colleagues to ban sunscreen products toxic to marine life. Perhaps all nano-lotions when washed off find their way to the ocean.

In the interim, until legislation is passed to ban life-killing substances, it's up to us as consumers to lessen the harm by taking a couple of minutes to read the label on a sunscreen before buying it. Sadly, oxybenzone and octinoxate are not the only nasties to be found in them. Any lotions and sprays that contain avobenzone, homosalate, octocrylene, or octisalate should also be left on the shelf. Those with zinc oxide and titanium dioxide are preferable as long as they don't contain nanoparticles (which they probably do if the sunscreen is clear). If the label reads "uncoated" zinc oxide, it has larger particles and is safe to use. Otherwise, try staying in the shade or under a beach umbrella, wearing a hat and loose cover-up clothing, and staying indoors during the hottest part of the day. Any of these remove or reduce the call for any type of cream.

If the Montreal Protocol can be adjusted to prevent more poisons from entering the sea or to stop the manufacture of deadly plastics, there are very likely other threats it could take on too. The search is on.

If one way of defending nature is to use existing law, another is to create a completely new kind of law that is tailor-made for the job—one with potentially far more power.

Making the Destruction of Ecosystems an International Crime

International law concerns interstate relations, responsibilities, and codes of behavior, while international criminal law is used for the most serious offenses against the global community. It can bring to justice individuals who are responsible for the very worst of crimes, to be tried in an international court and sentenced if found guilty. Four core international crimes, also called crimes against peace, were identified under the Rome Statute, the treaty that established the International Criminal Court. They are crimes against humanity, crimes of aggression, war crimes, and genocide. A growing number of people believe a fifth international crime must be added to the Rome Statute: ecocide, which is serious loss, damage, or destruction of ecosystems. British barrister Polly Higgins led the campaign, Stop Ecocide: Change the Law, to establish ecocide as an international crime in order to protect us from industrial practices that weaken or destroy the natural environment that we, and all living species, depend on. She asks, "Why haven't we criminalized mass damage and destruction to the Earth? We've normalized harm. We give permits to pollute. Whose interests are being protected here? The polluter—and not the community that is already impacted."

There are numerous UN resolutions, treaties, and development goals, but as Polly said, the harm carries on and escalates. International agreements like the Convention on Biodiversity and Paris Agreement are voluntary. Governments, business, and finance can't be forced to honor their commitments or be held accountable for the worst excesses of environmental damage. "The power of ecocide law is that it creates a legal duty of care that holds persons of superior responsibility to account in a criminal court of law."[9]

A short explanatory film asks us to imagine law as a pyramid with three layers (https://www.youtube.com/watch?v=ZvO4Dqwjdno). At the base is soft law, such as the Paris Agreement, which is unenforceable and doesn't protect us. The middle layer is civil law, where we find most environmental regulations, but suing companies is expensive and difficult. Corporations simply budget for lawsuits and fines and continue polluting as before. The top layer of the pyramid is criminal law, and only this can stop the harm.

A few years ago, when I was first exploring the idea of protecting the world's oceans in their entirety, Polly guided me through some points of law. She encouraged me to develop the concept further and believed it would fit well alongside ecocide law. I wanted to hear the ecocide story firsthand, so I went to see Polly in her campaign office, a large, bright room on the top floor of the old police station, looking out over the market town of Stroud, in southwest England. She welcomed me with a warm embrace, laughing about the ridiculous entry system that wouldn't let people into the building. We made the inevitable cups of tea, and after a brief catching up, I asked where and when it had all started and pressed "record" on my phone.

In 2009 Polly was onstage with the British journalist George Monbiot, speaking at the UN People's Climate Summit in Copenhagen. The discussion was thrown open, and someone in the audience said we needed new language to deal with mass habitat damage and destruction, and the loss of ecosystems.

> I said yes, it's like genocide, only it's ecocide. That should be a crime. It was a light bulb moment. I went home, dropped everything, and treated it like a legal brief. I studied and researched: What are the international crimes? What's already there, and how could this be made an international crime? That took me to the Rome Statute. I realized there was a missing fifth crime [that we needed] to criminalize ecosystem destruction. So I drafted a full proposal on how to amend the Statute to include ecocide as a crime and submitted it to the UN International Law Commission.

For a new international crime to be established, two-thirds of the member states must vote for it to be included in the Rome Statute (which currently equates to 82 of the 122 members, each state having one vote). When enough states approve an amendment, it becomes law. The campaign to make ecocide an international crime is widely supported by those who stand to lose most from the adverse impacts of climate change: communities living on islands or in low-lying coastal regions where they are particularly vulnerable to rising sea levels and extreme weather patterns. This is where Polly sought the support she needed, working to convince leaders, ministers, and officials that if enough of them cast a yes vote for ecocide to become an international crime, they will have the power to take on and beat the climate change perpetrators and safeguard their citizens in the long term. Some smaller states are financially stretched, and the fundraising arm of the ecocide campaign helps to meet the costs of sending a delegation to the annual meetings in order to cast their vote. The Republic of Nauru is a tiny island state in the southwestern Pacific. Its UN ambassador, Marlene Moses, expresses the mounting concerns of people living in island communities.

> For the people of small islands, understanding the importance of the ocean to human survival is as natural as breathing. If the ocean is healthy, we are healthy; if the future of the ocean is uncertain, so is ours. It is therefore alarming to witness the warming, rising, and acidification of the sea around us due to climate change, and the increasing pressure on tuna and other fish stocks we depend on, and it has made Small Island Developing States (SIDS) like mine among the loudest voices calling for international action.[10]

The surprise in the story came when Polly discovered that ecocide was indeed the missing international crime because it was originally included in the Rome Statute negotiations. It had been written into earlier drafts of the treaty and was even called ecocide. However, because of concerted, behind-the-scenes lobbying from certain industries, notably the pharmaceutical, agriculture, and fossil fuel sectors, the Netherlands, United States, United Kingdom, and France blocked the move, and ecocide was removed from the final draft. Now the legal principle is reemerging, with even greater purpose and potential, and the knowledge that ecocide was removed from the original legislation makes the argument to add it even stronger. "I'm just finishing a job that was started years ago," Polly said.

Not long after I visited her, Polly became ill with what she thought was a stubborn chest infection. After a few weeks when she was no better, further tests revealed that she had advanced and untreatable cancer. Polly died only five weeks after the diagnosis. I've known many environmental campaigners and conservationists over the years, but never one like Polly Higgins. She was an astute and determined strategist with a genuinely compassionate spirit and a childlike sense of fun—a curious amalgam of brains, passion, goodness, and humor. Polly was absolutely resolute about what she and her

team could achieve. It was refreshing and inspiring to see this optimism, and it was infectious. You couldn't help but feel more hopeful
about the future after talking to Polly Higgins.

Despite the shock of her unexpected death, Polly's campaign team,
led by her colleague Jojo Mehta, has been continuing her work, and
they are even more determined that ecocide law will one day be part
of the Rome Statute. Urging people to support the campaign, the website tells us, "there is nothing more powerful you can do with two minutes and fiver," and it is hard to argue with that.[11]

7: Worrying about the Wrong Stuff

I can't pinpoint any particular time or event as the start of my fascination with the sea. It was always there, with tooth fairies and ponies and snow. When I was growing up we lived on the coast near Liverpool in northwestern England. My father spent all his working life in shipping, and the house was adorned with seascapes and books about ships. On weekends and family vacations we would be either by the sea or on it, so I spent a lot of time playing in the surf with my brother and poking around in rock pools with a net, shivering in a damp towel and eating sandy sandwiches, or running across windy beaches with wet dogs. I was this little scrap of life beside a huge kingdom of liquid mystery stretching to the horizon, with Ireland and America somewhere on the other side.

States' not observing the rule of law is a serious hindrance to fixing broken oceans, but I believe we have a more profound issue to face—how we see ourselves in relation to nature and how most of us have come to regard the natural world as a whole. Societies have long used the sea as an inexhaustible larder and as a means of making money. When government and industry are propelled by those cultural values the results are devastating for wild places, because how we think determines what we do. The condition of coasts and oceans confirms this. Many corporations operate under the delusion that they are entitled to exploit nature without respecting it and to profit from it without taking responsibility for the damage it can cause. And corporate arrogance is frequently compounded by government collusion.

The Industrial Revolution "denatured" us. It pulled people away from the untamed and unpredictable realm of nature, and now we think of ourselves as separate from it. As Ian Johnston writes, "Some psychologists argue that urban industrialised living compromises an individual's sense of kinship with non-human nature, thereby opening the door to environmentally destructive behaviour. Simply put, humans don't protect what they don't know and value."[1]

We divide nature and believe we've conquered it. We chop it up into classes and orders, sections and subsections, carving oceans into bioregions, geo-regions, intertidal zones, and benthic realms, neatly separated into different elements as though they were unconnected and unrelated. The sea has been split politically and biologically, vertically and horizontally, though the wave that laps my toes on an English beach is part of the same water where blue whales feed off the shores of Sri Lanka. Coasts and seas are disordered, ragged and chaotic, and it seems incongruous to neatly categorize them in dealing with environmental issues. The result is a grand assortment of declarations and recommendations, of unattained targets, of neglected laws, directives, and resolutions made by international bodies, governments, and agencies. Departmental committees and panels of experts plan bilateral programs and regional initiatives. They produce evaluations and operational strategies, midrange targets, and quantitative risk indexes. They organize seminars and forums and meetings about meetings, recommending integrated adaptive marine spatial planning, ocean zoning, predictive habitat modeling, and analysis of the compatibility of ecosystem services. Certainly all this effort has made some headway, but overall it hasn't kept coasts, seas, and oceans anywhere nearly safe enough from harm by humans. Authorities become aware of problems as isolated issues once the damage has been done. They attempt to tackle each one with a new policy or regulation, usually too weak and too late be effective. It's a reactive approach when we need a proactive one: one that recognizes the natural environment as one elaborate system whose parts are linked and inseparable.

During my research I often heard and read the term "ocean management." It's widely used by marine conservation organizations, in academic journals, and in government reports. There's something revealing about this. The concept of people managing oceans is a peculiar one. Modern humans have been around for about 200,000 years, while the oceans have been around almost as long as the Earth itself. They were formed an estimated 3.8 billion years ago, which makes oceans 19,000 times as old as humans. Considering that language helps to shape our thoughts, the term "ocean management" implies that we control the sea—Earth's largest and wildest natural system— and that without us it will falter. In fact the contrary is true. We cannot manage oceans; they have managed themselves unaided for millions of years. Instead we should be managing ourselves in a way that safeguards nature's largesse. "Ocean management" then becomes "people management."

Pioneering Philippine lawyer Tony Oposa Jr. stresses the need to change terms in order to change attitudes. In his TED X talk recorded in Manila, he examines the word environment, for instance. With an impassioned gesture of the hand he says emphatically, "The difficulty with the word environment is we don't understand it with the mind and the heart. It is not about the birds and the bees and the flowers and the trees. It is about life! And the sources of life!"[2] He goes on to describe the astounding natural wealth of his homeland and how the seas around the Philippines are the most biodiverse waters on Earth. He explains that scientists have located the center of marine biodiversity in the Sulu Sulawesi Marine Triangle, a region lying between Indonesia, Malaysia, and the Philippines. Further studies show that the center of this center of biodiversity is the central Philippine islands. Then his voice drops: "But look at what we've done," he goes on, "using the paradigm of consumption and destruction, in a matter of a hundred years. Out of the three million hectares of coral reefs in the Philippines barely fifty years ago, we now have less than 1 percent that is intact. If the economics of extraction and consumption has got us into the problems we are in now, then the opposite must get us out of them—which is conservation, protection, and restoration."

With marine reserves and better fisheries regimes, with the legal framework for universal marine protection, with grassroots counteraction, and global reach monitoring and surveillance, all the solutions to rescuing and conserving the world's oceans are there, ready to go. But governments and their leaders are prioritizing the wrong stuff. And all too often the media are their sidekicks, distracting us with flimflam and urging us to care about relative trivia. The importance of news is conveyed by its prominence and by the amount of air time or newspaper coverage it takes up, which gives news editors considerable influence over the public's consciousness. They can subtly manipulate our values and our worldview simply by deciding on the pecking order of news items. One afternoon I bought a few groceries and a newspaper, and as I walked out of the store I read a front-page story that was truly disturbing. It carried the headline "Biological Annihilation: Earth Faces Sixth Mass Extinction." The article was based on a study published in the American journal *Proceedings of the National Academy of Sciences*, which found that populations of thousands of species of birds, amphibians, mammals, and reptiles are in decline, largely owing to human activities. In their introduction, the study authors Gerardo Ceballos, Paul R. Ehrlich, and Rodolfo Dirzo write,

> Dwindling population sizes and range shrinkages amount to a massive anthropogenic erosion of biodiversity and of the ecosystem services essential to civilization. This "biological annihilation" underlines the seriousness for humanity of Earth's ongoing sixth mass extinction event.[3]

"Ocean temperatures hit record high as rate of heating accelerates. Oceans are clearest measure of climate crisis as they absorb 90% of heat trapped by greenhouse gases" was the headline some months later.[4] The sixth mass extinction is upon us and the oceans are warming up dangerously fast, yet the reaction in the corridors of power is minimal, even with this most alarming information. And which is the most disturbing truth? The depressing results of this research? Or that the report triggered little political response? Or that in the mainstream media such earth-shattering news is regarded as less important than what the Duchess of Cambridge was wearing and the result of a third-round match at Wimbledon? Because the day I first read about the sixth mass extinction on Earth it was outranked by those news stories. What's more, in several newspapers, television broadcasts, and radio bulletins it didn't feature at all.

Despite much of the media's confused sense of what matters, pretty much everyone is aware of the major environmental troubles afoot, on land and in the sea. At present, destructive commercial fishing and plastic pollution are widely considered the most acute problems to tackle in the sea, but potentially bigger problems loom and lesser ones lurk. When I typed "threats to the ocean" into a search engine, I saw that most were specifically quantified: "the five worst threats to our oceans" or "the top ten problems facing the sea today." There are four, or six, or seven perils and ten, twenty, and even twenty-five threats to the ocean to deal with, depending on which website or article you read. I was bombarded by lists of risks and perils, dangers and issues and threats. They blur into one giant mess that we need to sort out. I see them not as disconnected problems but as parts of a complex web of damage and decline, forming its own sprawling ecosystem of troubles. Habitat destruction and biodiversity loss; rampant industrial fishing; corruption, and labor exploitation; plastic debris and discarded fishing gear; 90 percent of top predators gone; all kinds of pollution; damage or noise disturbance from shipping, sonar, construction, oil and gas exploration and mining; an inadequate system of governance and law enforcement; and the effects of climate change—which include ocean acidification, coral bleaching, unpredictable and extreme weather events, warming seas, melting ice, and rising sea levels. In my mind the problems and threats cluster into a hideous watery mass that seems insurmountable. Sometimes, in fact most times, perhaps childishly, I want everything to be so very much *not* the way it is. I wish it were all a bad dream and that everything in the ocean were completely fine.

The Plastic Curse

Occasionally there are dramatic accidents that contaminate the sea on a grand scale, such as Japan's Fukushima nuclear plant meltdown and the *Exxon Valdez* tanker spilling its gruesome load of crude oil into Alaska's Prince William Sound. Meanwhile, the long-term, cumulative effects of pollutants entering the sea every day also present a serious problem. Significant quantities of waste and pollution stem from shipping and the oil, gas, and mining industries, but the greatest proportion originate on land (approximately 80 percent): agricultural runoff (which causes dead zones),[5] industrial discharges, raw sewage, oil leaks, discarded fishing gear, wet wipes, toothbrushes, cigarette butts, residual pesticides, and antibiotics from fish farms—you name it—it's in the sea. The most notorious ocean affliction of the times, however, is plastic waste.

Oh, how we fell for plastic. It was the wonder material of our shiny modern world. Plastic was a marvel; it was light and strong and was cheap to produce. Mostly it was durable; you could rely on plastic. We welcomed it into almost every aspect of our lives: we drink from it, eat from it, and bathe in it. Plastic carries our belongings, plays our music, makes our children's toys and a good part of our household goods, clothes, cars, and computers, what we sit on, and often what we sleep on. We wash with plastic, we wear plastic, and sometimes we even eat it, because when some of the fish on our plate swam in the sea they mistook tiny microbeads floating in the water for food. But as with so many love affairs, ours with plastic is turning sour. And as often happens with ex-lovers, those qualities we once so admired became the very qualities we began to resent and grew to detest. Durability was considered a good thing until we discovered that plastic sticks around for hundreds or maybe thousands of years. That's not durable, that's everlasting. And even when we came to see that plastic was bad for our world and understood it was time to be free of it and move on to benign alternatives, we couldn't give it up and kept coming back for more. The plastic peril had infiltrated society and settled in, to the extent that we can barely manage without it and now it seems almost impossible to get rid of.

Then we discovered that plastic doesn't remain in its visible form, as a bottle, a bag, a pouch, or a pot, perhaps. Instead, it breaks down into millions of tiny particles that can't be gathered up once they've dispersed onto the land or into the ocean. And they go everywhere, even into the bodies of tiny creatures from the deepest seas. Researchers at the Scottish Association for Marine Science in Oban, Scotland, found that 48 percent of the animals they tested living in the Rockall Trough, to the west of the British Isles, had ingested plastic particles. "When plastic trash degrades in the ocean, it doesn't just go away: It becomes countless tiny particles, and little creatures called larvaceans sweep it up—and into the food chain."[6]

Plastic is everywhere in the ocean. It is in Arctic sea ice and right down in the deepest part of the ocean, on the floor of the Mariana Trench. It pollutes rivers, fouls beaches, and is poisoning and choking untold numbers of beautiful wild creatures. Over the decades since it sneaked into our lives, our once innocuous plastic friend has ballooned into an uncontrollable and lethal menace despite international law to prevent marine pollution. The relevant agreements are the London Convention on the Prevention of Marine Pollution by Dumping of Wastes and Other Matter (the subsequent London Protocol was agreed on to bolster the original treaty), and the International Convention for the Prevention of Pollution from Ships, also known as MARPOL. Plastic statistics are chilling. An estimated 8.8 million tons of plastic waste reaches the ocean each year,[7] and that amount is projected to double by 2030 and quadruple by 2050 unless there is a radical shift in present systems and practices. Sadly, what we see bobbing on the surface or littering the coast is a small proportion of it, because most marine plastic waste is moseying around the water column or lying on the seafloor.

Across the world, regional, national, and international strategies to tackle the marine plastic curse are coming thick and fast, too numerous to describe here. For example, being second only to China as the largest contributor to the problem, Indonesia has ambitious plans to reduce plastic waste in the sea by 70 percent by 2025. (A detailed study published in the American journal *Science* found that up to 60 percent of plastic waste in the sea was coming from just five countries: China, Indonesia, the Philippines, Vietnam, and Thailand.)[8] Scientists and policy makers from Ghana, the United Kingdom, Vanuatu, New Zealand, and Sri Lanka make up the Commonwealth Clean Ocean Alliance, a cooperative effort aimed at eliminating single-use plastic and tackling plastic waste in the sea.[9] Tens of millions of plastic drinking straws are used and discarded every day, and millions end up in the water or on beaches. As a result, the number of bans on them are increasing—starting small in cafés, museums, schools, and offices and extending to cities and states, such as those in Seattle, Vancouver, and California, and to nationwide bans as in Canada, which is planning to see the end of single-use plastics by 2021. The plastic counterattack needs three essential elements: accountability by industry and manufacturing; public awareness and personal responsibility; and governments' passing protective laws and funding the development of alternatives to plastic. A cross-sector, cross-border counteroffensive on the proliferation of plastic waste could stop most of it at the source, and in time biodegradable and benign alternatives could become the norm.

Additional Threats to the Oceans

During my online search for the bad stuff, three less well-known types of activity cropped up repeatedly that I felt needed some explanatory backstory: deep-sea mining, marine fish farming (marine aquaculture), and ghost fishing.

Deep-Sea Mining

Offshore mineral, oil, and gas industries have been environmentally controversial for decades. Drilling for oil or gas and transporting it across the sea is a risky operation. When things go wrong they go really wrong, as with the explosion of the *Deepwater Horizon* rig in the Gulf of Mexico, when eleven people were killed and over 200 million gallons of crude oil escaped into the sea. Even when things don't go wrong, exploration, drilling, laying pipes, and transporting oil and gas are damaging. There are regular leaks of oil and lubricants from wellheads and pipelines, plus discharges of oily ballast water from ships. The industry considers it acceptable to expect these "minor" day-to-day mishaps on and around rigs, but it all mounts up, and hundreds of thousands of gallons of oil "inadvertently" enter the world's oceans each year.

Another threat that has emerged more recently is deep-sea mining to retrieve mineral deposits from the seabed and beneath it, including copper, nickel, aluminum, manganese, zinc, lithium, and cobalt. Deep-sea mining has many of the same detrimental effects as oil and gas extraction, plus others. Machines built for mining the seabed are sinister beasts, like something from a science fiction thriller. Giant excavating robots built for Nautilus Minerals Inc. are ominously called seafloor production tools. There are three types: the auxiliary cutter, the bulk cutter, and the collecting machine, each about fifty feet long by fifteen to twenty feet wide and weighing 242 to 374 tons. The audacious intention of Nautilus is summed up thus: "Nautilus Minerals believes the future of mining lies at the bottom of the ocean. The Australian-Canadian company . . . revealed some of the remote-controlled vehicles that will be cutting up the ocean floor in search of copper, gold, nickel, and other minerals."[10]

Note the forbidding phrase "that will be cutting up the ocean floor," meaning that operating these machines clearly damages and destroys deep seabed habitats that support rare and unique species in unspoiled seas. Besides removing large areas of the seabed, deep-sea mining sends up clouds of sediment that then settle and smother habitats. The noise and vibration are extremely disturbing and can be fatal to wild species, and toxic pollutants will inevitably leak from the machinery into the water. In addition, deep-sea mining disrupts and releases carbon stored in seafloor sediments, exacerbating the effects of climate change. A paper published in 2019, coauthored by a number of marine scientists led by Dan Laffoley, spells out eight urgent actions needed to stave off the ecological breakdown of the oceans. On the issue of deep-sea mining, they are concerned because the International Seabed Authority has already granted twenty-nine exploration licenses for commercial mining in an area estimated at over 1,540,000 square miles while knowing little about the life within those areas. The scientists call for a precautionary moratorium on deep-sea mining:

There has been a rapid increase in the number of countries and companies seeking exploration access to the ocean floor. . . . Meanwhile, scientific surveys conducted in prospective mining regions (ferro-manganese encrusted seamounts, polymetallic nodules on abyssal plains and seafloor massive sulfides at hydrothermal vents) have confirmed hundreds of new species, as well as high diversity in both species and habitats. Many of these areas are considered vulnerable marine ecosystems, in that they are structurally complex, and contain endemic, rare, long-lived, slow-growing and fragile species. Deep sea mining will add to the stressors already facing the ocean, and probably lead to cumulative impacts, which will further undermine ocean health and resilience.[11]

The International Seabed Authority has so far not turned down any application for exploratory mining. According to Greenpeace, "Created in 1994, the ISA is meant to organise and regulate deep sea mining activities in the international seabed, but far from protecting our oceans, they are selling it off to greedy industries that are trying to plunder our ocean floor for profit."[12]

Marine Fish Farming (Marine Aquaculture)

As wild fish stocks have fallen because of overfishing, farmed fish have become the world's fastest-growing source of animal protein. Species raised in marine fish farms include sea bass, cod, mussels, oysters, salmon, prawns, and shrimp. Marine fish farming is generally intensive and can have significant negative effects on surrounding natural environments and wildlife. The damage done varies depending on what is farmed, which methods are used, and where the farm is located. Warm-water prawn and shrimp farming in Southeast Asia, for instance, does serious harm to coastal habitats, particularly mangrove forests. (See chapter 9 for more on the damage caused by prawn and shrimp farming.) Salmon farming is a very big player in the industry (Atlantic salmon is the dominant species), with about 2.75 million tons of farmed salmon produced annually.[13] The salmon are reared in large pens called net cages placed in the open sea. The fish live in unnaturally crowded conditions and are vulnerable to disease, so they are regularly dosed with antibiotics and pesticides. Surrounding undersea habitats are affected, and some wild species become sick or die because waters are contaminated with feces, uneaten food pellets, parasitic lice, dead fish, and chemical residues flowing out of the net cages. Sea lice are a special problem, since they grow resistant to lice treatments as fish farmers add higher doses or stronger pesticides to the water, and these substances are entering wild habitats. As Don Staniford, the director of Scottish Salmon Watch, put it, "What we are seeing now is a chemical arms race in the seas."[14]

In response to criticism about high levels of pesticides, some fish farming companies are opting for what they call the "more environmentally friendly and natural" remedy of using cleaner fish called wrasses. These natural predators of lice will happily swim among the salmon eating away their parasites. But wrasses are far from being an environmentally friendly solution because, until the industry can breed enough of them, they are using wild-caught fish and depleting those natural populations.

Farmed salmon are generally fed pellets made of a mix of wild fish, ground-up chicken feathers, genetically modified yeast, soybeans, chicken fat, and a protein powder called ethoxyquin, developed by Monsanto in the 1950s as a pesticide. The wild fish part of the salmon's diet frequently comes from hundreds or thousands of miles away, such as anchovies from Peru and Chile brought over for salmon farms in Norway and Scotland. Catching wild fish to feed the caged fish can exhaust wild stocks, which has a knock-on effect right along the food chain, depriving other marine species of food. The Scottish salmon farming industry plans to expand, and a study by the organization Feedback (which campaigns for sustainable ways of producing food) found that, in order to fuel the proposed expansion,

the industry will need to increase its use of wild fish from around 460,000 tonnes a year, to 770,000 tonnes a year. For context, the current quantity of wild fish fed to farmed Scottish salmon is roughly equivalent to the amount purchased by the entire UK population, and to fulfil growth ambitions this amount would need to increase by approximately two thirds. In addition, the industry relies on other land-based feed ingredients, such as soya and palm oil, which present widely-documented sustainability challenges in themselves.[15]

Significant numbers of farmed salmon escape into the surrounding sea, risking the viability of the wild population if the fish interbreed. A farmed salmon, Paul Greenberg writes, has been bred to "eat a lot and grow fast . . . but it has lost many of the fierce, determined traits that make a wild salmon able to swim against powerful currents, withstand fluctuations in temperature and spawn in a river besieged by predators."[16]

Marine mammals fall afoul of marine fish farming too. The net cages attract fish-eating predators like seals, dolphins, and sea lions. Seeing them as a threat to the caged stock, some fish farmers shoot the animals. Others drown when they get entangled in the nets surrounding the cages. Few consumers are aware that marine mammals may have died in the production of their salmon steak.

Even if you care little about the ecological concerns surrounding farmed salmon, there is another very good reason to avoid it—your own health. The cocktail of unpalatable ingredients that make up the fish feed are transferred, to some extent, into the body of whoever eats the salmon. And there's more: the flesh of farmed salmon is a dreary gray, so a dye called canthaxanthin (E161g) is added to the feed to turn the flesh pink, making the end product more appealing to customers when displayed in supermarket cases and on fishmongers' slabs. Farmers can even choose precisely which shade of pink they'd like from a color chart known as the Salmofan (created by the pharmaceutical giant Hoffmann-LaRoche). Just like one you might use to select the right shade of paint for your kitchen or bedroom, the Salmofan opens out to reveal a range of pinks, from very pale to deep rose. The Australian and New Zealand authorities have banned the use of canthaxanthin in the food industry because it might damage the retina.

Farmed salmon has been described as the most toxic food in the world.[17] Even back in 2004, a study published in the journal *Science* confirmed that concentrations of cancer-causing substances were present in European farmed salmon (from Norway, the Faroe Islands, and Scotland) and advised readers to limit consumption to no more than one meal every four months to reduce their risk of cancer.[18]

Don Staniford likens more and more salmon cages to "a malignant cancer on our coasts": "Everywhere this dirty rotten industry operates there are problems with mass mortalities, infectious diseases, waste pollution, toxic chemicals and impacts on marine life. Salmon farming is a pariah not a panacea. By cramming the 'King of Fish' into factory conditions we have turned *Salmo salar* (Latin for 'The Leaper') into a leper."[19]

Some years ago I talked to a scuba diver who worked for a salmon farm in the Republic of Ireland. His job was to check that everything was all right in the large net cages suspended in the sea just off the craggy coast of Kerry. He would swim down into each one to collect any dead fish, make sure the cages weren't damaged, and so on. "You must get plenty of lovely salmon as a perk of the job, I suppose?," I asked him. His response took me by surprise: "I would never touch the stuff, and if you'd been down there and seen those fish, bunched together, deformed, and full of chemicals, you wouldn't either."

There is broad agreement in the conservation movement that current methods of farming salmon should be reformed and that using net cages in the open sea has to end. Moving to a land-based closed containment system would create a barrier between wild species and the farmed fish. When I see salmon in the supermarket, I'm surprised by how inexpensive it is. If the fish cost the customer more, farming could be made less intensive, fewer chemicals would be necessary to offset problems caused by overcrowded pens, and farmed salmon would be a healthier food. In Don Staniford's words, "There's no right way to do the wrong thing. Just stop."

Ghost Fishing

For two months, scientist Chris Warren was a volunteer crew member on the former US Coast Guard Island-class cutter the *Farley Mowat*. It is one of several vessels in the fleet of the conservation organization Sea Shepherd.[20] In early summer 2017 the ship was patrolling the Vaquita Refuge in the Gulf of California as part of the Milagro III cam-

paign to defend the highly endangered Vaquita porpoise. Chris wrote
to me about what they came across one day:

> We found a leatherback turtle floating in the Gulf of California Vaquita
> Refuge. At about one and a half metres in length, drifting motionless
> on the surface in near still waters, we spotted it from afar. We drew
> closer and we could see the net around it. The turtle was completely
> entangled and had been dead for some time, and the net stretched
> down to two anchors in the seabed a few hundred metres either side.
> This beautiful creature was completely alone in a vast sea, reduced
> to a floating corpse, still trapped even in death and blown up like a
> balloon.

Ghost fishing is what discarded fishing gear does: it continues to
trap and kill sea creatures long after it was last deployed. Traps, pots,
nets, hooks, and longlines either are lost or are deliberately thrown
into the sea or left on beaches, and they carry on killing, sometimes
for years. Fish, crustaceans, sharks, rays, mammals, birds, turtles—
all kinds of sea life become entangled in the ghost gear and die by the
millions from exhaustion, suffocation, starvation, or drowning. The
nets also damage live coral and smother reefs. Even the mighty blue
whales are in danger. In 2016 an eighty-foot blue whale was spotted
off the coast of Southern California tangled in hundreds of feet of fishing line with buoys attached. Rescuers were very concerned that the
whale was tiring and might eventually die, but they were unable to
reach the animal and remove the line. Stressed by the attempts of its
would-be rescuers, the whale dived and couldn't be traced.

Abandoned fishing gear is a deadly menace worldwide. The international animal welfare organization World Animal Protection reports that an estimated 704,000 tons of ghost gear is left in the world's
oceans every year, inflicting incalculable cruelty and death on marine
life. Retrieving discarded gear is one way to tackle ghost fishing. Chris
Warren described how he felt when, with some difficulty, he and his
teammates pulled an abandoned net out of the water onto the deck
of the *Farley Mowat*.

Whilst we pulled up the net I felt physically and emotionally exhausted. Anger towards those responsible fuelled my efforts, but they will never know the damage they cause. Completing the net haul brought complete elation, the feeling that although some died, many more marine lives may have been saved and that such a challenge was executed so perfectly by a group of passionate volunteers. The scale of the devastation caused by these nets is incomprehensible. A barren ocean, which should be teeming with whales, dolphins and turtles, instead resembles a desert under the blistering sun. Whilst we do all we can, we know that it will never be enough to reverse the impact of others. Every life saved gives purpose to our own and makes us feel a little less ashamed.

Nets used to be made of natural and biodegradable materials like hemp, coconut fiber, bamboo, or cotton but with the coming of synthetic materials, nets no longer rot in the water. To learn about ghost fishing is especially distressing, perhaps because of the scale of the suffering it causes, coupled with the disturbing fact that it is almost entirely avoidable. There are moves to tackle this insidious killer, but it is like facing a mountain to climb, and we are still in the foothills. In 2015 the Global Ghost Gear Initiative (GGGI; https://www.ghostgear .org/) was launched. It is a cross-sector alliance bringing together participants from fifty organizations—from academia, industry, government, and conservation organizations—to work together and find ways to address ghost fishing. As with nearly all of these issues, there is no one-stop solution. Finding ways to end ghost fishing calls for a multipart strategy, and there are some very inspiring and inventive projects combating ghost fishing in many parts of the world.

Danajon Bank is a special place: a double-layered coral reef, one of only six in the world, extending about a hundred miles between the islands of Bohol, Cebu, and Leyte in the central Philippines. It is regarded as the king of kings in wealth of ocean life because, as Tony Oposa said, it is "at the center of the center" of what scientists consider the most biodiverse marine environment of all, the region where they believe almost all Pacific marine life evolved. However, with fifty island communities around the reef almost entirely dependent on the sea for their livelihoods, Danajon Bank is under severe population pressure. It is overfished and polluted and has become one of the most degraded coral reefs in the world. Fishers had been discarding their gear into the sea or on the beach for many years, but now the situation is changing. Bohol resident Isoy dives down to bring up old nets lying on the seabed or snagged on the coral. He and his neighbors bundle the collected nets, which are weighed at a nearby depot and then sold to Net-Works (http://net-works.com/), a collaborative initiative between the conservation charity the Zoological Society of London, carpet manufacturer Interface, and recycled yarn producer Aquafil. Net-Works collects unwanted nets and lines from island communities around Danajon Bank and recycles them to make carpet tiles. People no longer abandon nets overboard; they sell them to be recycled. They also collect and sell other nets and lines left in the sea or on the beach. The extra income is especially welcome, because declining catches leave less money from fishing. Together with the project in Cameroon, Net-Works has collected 246 tons of waste fishing nets for recycling since 2012 and helped 2,200 families financially. It's a fine example of the circular economy in practice, with a range of positive outcomes: the people of the islands' coastal towns and villages are better off and able to spend more on health and education or to invest in new enterprises such as seaweed farming; choked seas and beaches are being cleared of nets, allowing marine environments to recover and fish to return; and Interface can make carpets with the waste material. The enterprise has been so successful that ambitious plans are brewing to scale it up and set up programs farther afield.

Similar projects are running in other parts of the world. The Healthy Seas Initiative retrieves abandoned nets, primarily in European seas, and regenerates them into high-quality yarn to make sportswear. "Nets to Energy" in Hawaii uses discarded nets to generate electricity, and in Chile an organization called Bureo collects waste nets from the ocean and turns them into—I wasn't expecting this—skateboards.

8: The Silver Bullet?

Creating marine protected areas and marine reserves is generally believed to be an ocean conservation silver bullet. Every year more protected areas are designated, steadily leaving less and less water exposed to misuse, so in a sense the transition to seamless ocean protection is gradually under way already.

The world's first official marine reserve, Cape Rodney-Okakari Point Marine Reserve, better known as Goat Island Reserve, is a snippet of inshore water on the eastern coast of New Zealand's North Island. It was officially protected after a lengthy campaign led by botanist Val Chapman, zoologist John Morton, and marine biologist Bill Ballantine. They wanted a short stretch of coastal water cordoned off from human disturbance as a "sea laboratory" for studying resident species in their natural state. After years of lobbying, New Zealand's Marine Reserves Act was passed in 1971, making it possible to set up lawful reserves. A long tussle with the authorities ensued, and finally in 1975 the Goat Island Reserve was established. Covering a little over two square miles, what the reserve may lack in size, it easily makes up in impact.

By the 1970s overfishing of snapper and crayfish in and around Goat Island waters had enabled sea urchin numbers to escalate. Without their main predators to keep the population in check, urchins grazed the reef kelp down to barren stubble. After 1975, with the new reserve in place and fishing stopped, snappers and crayfish came back, restoring the balance of predator and prey. Urchin numbers fell, the kelp grew back, and many other species returned, like blue maomao and marblefish. Catches for local fishers swelled as replenished stocks moved beyond the protected area. Bill Ballantine and his colleagues were doubly surprised by the reserve's returns. They didn't expect undersea flora and fauna to rebound so swiftly, nor did they foresee the surge of public enthusiasm for this oasis of ocean life. Throughout the year thousands of people visit: groups of schoolchildren, hikers, families on day trips, snorkelers, and divers go there to enjoy a profusion of marine life along the shore and in the water. And in an ironic and telling turn of events, many of those who had strongly opposed the reserve became stalwart supporters. Bill realized the healing potential of protecting even small areas of the sea, and until his death in 2015 he became an active reserves campaigner, traveling the world to speak at public events and talking to scientists and policy makers about how crucial reserves are for safeguarding marine biodiversity. In New Zealand's waters there are now over one hundred protected areas of all sizes, from the pint-sized, trailblazing Goat Island Reserve to the immense Kermadec Ocean Sanctuary. Surrounding a string of uninhabited subtropical islands 560 miles northeast of New Zealand, Kermadec spans 240,000 square miles (more than twice the area of Italy). The five islands are peaks in a chain of submerged volcanoes rising six miles to pierce the ocean surface. The sanctuary contains many undersea habitats and is home to a huge variety of life: thirty-nine species of seabirds, thirty-five species of whales and dolphins, and a wide variety of other sea life, including fish, corals, and shellfish found only in these waters.

From the oldest reserve and the largest reserve, we go back to November 1769 to learn about what is perhaps New Zealand's most biodiverse reserve relative to its size. From the deck of the British Royal Navy's research ship HMS *Endeavour*, sailing along the northeast coast of New Zealand, Lieutenant James Cook spotted a group of little islands. These disorderly dollops of rock in the sea reminded him of a popular dessert of the time, poor knights' pudding, hence their peculiar name, the Poor Knights Islands. Here temperate and subtropical waters merge, creating an especially rich marine ecosystem. Below the surface, the remains of an ancient volcano have left an underwater landscape of caves, arches, tunnels, and cliffs, which are home to a great range of marine plants and animals. Masses of blennies, gobies, triplefins, and pink, orange, and yellow anemones live among soft corals, sponge gardens, coral fans, and kelp. Swirling shoals of kingfish, yellow snappers, white trevally, pink maomaos, and blue maomaos cruise the shallows and the depths. There are green wrasses, pig fish, lizard fish, mosaic moray eels, John Dories, and thousands of two-spot demoiselles. It's a riot of fish, with over 125 species found here, plus all sorts of other marine life: shellfish, snails, starfish, and urchins. With so much diversity, half a mile of the sea surrounding the Poor Knights Islands was made New Zealand's second reserve in 1981, with full protection secured in 1998. A good many larger animals can be seen in and around the reserve as well, including stingrays, dolphins, seals, turtles, and several types of whales. Numerous species of birds come to the islands to raise their young, the most abundant being Buller's shearwaters, which nest in burrows. The islands are particularly special to these birds because it is the only place in the world where they breed.

The Poor Knights Reserve is the country's most popular location for scuba diving and snorkeling, and for many New Zealanders it is the jewel in an undersea crown. Adventure travel writer Charli Moore described the reserve for her online travel guide Wanderlusters:

> With such a diverse underwater environment there is no wonder that it is recognized as one of the few remaining pristine eco systems on the planet.
>
> The lack of human interference here has encouraged the growth of a subtropical playground for marine and plant life, with thousands of different species taking refuge here throughout the year divers experience an underwater world like no other. . . .
>
> The dark waters made you feel as though civilisation had rewound itself and you were diving beneath the gaze of the ancient Maori tribes who once inhabited the land above. . . .
>
> The wealth of life was astounding.[1]

Those protecting an area of sea usually aim at conserving a particular habitat or an important breeding ground for an endangered species. When waters are left unfished and undisturbed, sea life can mature, breed, and become plentiful again, and the recovery of habitats and depleted populations of wildlife can be remarkable. Numbers of mutton snappers, for instance, multiplied by fifteen in only four years in Tortugas Ecological Reserve in Florida;[2] scallops increased fourteenfold after five years of protection in a New England fishery;[3] and in Spain's Cabo de Palos, the population of dusky groupers grew more than fortyfold in eleven years.[4] In 2009 a group of scientists led by Sarah Lester published a study on the biological changes in 124 marine reserves.[5] On average, biomass (the mass of animals and plants) more than quadrupled over three years. The upsurge of life spilling over the reserve boundary also increases the catch of fisheries in the surrounding area.

There are several names for these protected areas. They may be called national parks, marine reserves, marine protected areas, marine parks, blue parks, marine monuments, marine conservation zones, marine sanctuaries, marine nature reserves, sites of special scientific interest, specially protected areas, special areas of conservation, or ocean sanctuaries. Differing terms for the same thing are confusing, but essentially there are two types of designations; a marine reserve is fully protected, with commercial activities like fishing or dredging permanently prohibited (also called "no-take" zones), while a marine protected area or MPA has some limitation on activities (but what those are differs considerably). Most MPAs are open for recreation and some fishing or other commercial uses (known as multiple use). Some have only seasonal restrictions, and other restrictions cover a particular species or element within the area. (Strangely, the authority of the Marianas Trench National Marine Monument, for instance, covers the submerged lands but not all of the water above.)

Establishing a reserve or marine protected area can be tortuous, fraught with bureaucratic obstacles. There is always opposition from various sectors such as the oil, gas, and aggregate dredging industries, or perhaps from recreational fishers and angling associations. Most commonly it comes from those in the fishing industry—at least in the beginning. The story of a particular protected area in the United Kingdom illustrates this very well.

Regenerating Lyme Bay

Lyme Bay is a generous bite out of England's south coast where the counties of Devon and Dorset meet. It is Thomas Hardy country, the windswept hills making a moody and mysterious setting for his novels. Standing on the cliff near the village of Eype, I looked over a sea gilded with reflected rays of autumn sun and tried to picture what lay beneath. The waters of the bay are home to about three hundred recorded species of plants and animals: sponges, cuttlefish, sea fans, starfish, and rare sunset corals. The presence of the pink sea fan *Eunicella verrucosa* indicates that this is a complex and delicate ecosystem.

To protect this special area from intensive fishing, in 2008 the government's Department for Environment, Food and Rural Affairs introduced a permanent ban on scallop dredging and trawling in sixty square miles of the bay. Hilary Benn, environment secretary at the time, said, "The environmental benefits will be huge and species under threat will be able to recover and thrive." Although the news was welcomed by conservation groups, local fishers were up in arms about the announcement. After the government ruling was imposed, however, a major flaw in the plan became apparent. The ban on destructive fishing practices was for mobile fishing gear and didn't include static gear. So, although trawling and dredging had to stop, the use of pots and static nets doubled within the protected area. Overfishing continued because people just used other methods more. As so often happens, the conservationists and the fishers were at loggerheads, and the usual opponents were back in the ring together: protection versus exploitation. Then entered a referee in the person of local conservationist and diver Neville Copperthwaite. The ocean conservation organization Blue Marine Foundation (BLUE) became involved and asked him to mediate. I arranged to meet Neville in West Bay, one of the fishing ports in Lyme Bay, and he told me the story.

There are four small fishing ports within Lyme Bay's protected area: West Bay, Beer, Lyme Regis, and Axmouth, and fifty-nine fishers were operating forty-five boats in the bay. In autumn 2011 Neville invited them all to a meeting to discuss the dispute, hoping to build some bridges. The men were wary and reluctant to attend the first meeting, and some refused to go. "The first meeting was frosty," Neville said. "We were in a cafe in West Bay. There were twenty to thirty fishermen versus myself, Tim Glover and Charles Clover from BLUE. We listened to them and digested their concerns. One of the biggest challenges was getting them to trust us."

The fishers had to be convinced that a ban could bring positive outcomes. Little by little they were coming around and were invited to take part in planning a regime that would safeguard the fishery for the long term. After further meetings and with greater communication and understanding, mistrust gradually subsided. Ultimately both parties wanted the same thing for the bay, they just had to realize that and work together. "Success is more likely when fishers have shaped the rules," Neville concluded. In time a voluntary code of conduct was agreed on by everyone, including both commercial and recreational fishers. Measures included fitting each boat with a vessel monitoring system (VMS) and restricting the size and amount of fishing gear (for example, capping the number of lobster and whelk pots each vessel uses).

Judicious fishing in Lyme Bay Fisheries and Conservation Reserve has certainly paid off, with habitats recovering and stocks replenished (reef species of flora and fauna, for instance, have quadrupled since 2008).[6] The reserve management team has also set up a fish brand, "Reserve Seafood." Through the scheme, Lyme Bay fishers are now supplying customers with fish and shellfish caught responsibly from well-managed stocks, and they are fetching a market premium of 25 to 30 percent. Top London restaurants can't get enough. Chiller units on the harborsides keep fish cool and fresh from the moment they are landed, ready to be delivered to markets and restaurants far afield. Now most people agree that the whole initiative is the best thing to happen to these ports for a hundred years.

Clearly the Lyme Bay experiment is doing well. Creating a voluntary code and an action plan that is debated and agreed to by all stakeholders—fishers, conservationists, scientists, and local authorities—is regarded as a model of self-management for responsible continued fishing. And as happened in New Zealand when Bill Ballantine set up the Goat Island Reserve, those who were originally vociferous critics later became enthusiastic participants and the reserve's staunchest defenders.

Thinking Big

Across the world, although there are many areas with a degree of environmental protection, most are small: according to Protect Planet Ocean, only two square miles or less. However, the number of protected areas is growing—and the sizes too—some cover several hundred thousand square miles. These are called very large marine protected areas (VLMPAs) or large-scale MPAs, defined as those covering about 40,000 square miles or more. Over the past few years a kind of international contest has developed about who has the biggest marine protected area and who cares most about the sea. "We've got the biggest MPA in the world," announces one country, "OK, then we'll make an even bigger one," announces another. It seemed to kick off with Bill Clinton's executive order establishing the Northwestern Hawaiian Islands Coral Reef Ecosystem Reserve in December 2000 during his final days as president—a parting gift for oceans. It covered 131,000 square miles and was subsequently incorporated into the larger Papahānaumokuākea Marine National Monument. In the following years there was a sequence of official announcements by other countries designating more expanses of protected ocean.

In 2014 President Barack Obama's executive order created what was then the world's biggest marine protected area. He authorized the expansion of an existing reserve to six times its original size. The Pacific Remote Islands Marine National Monument consists of seven uninhabited islands and atolls under American control. The islands are dispersed in the central Pacific with extensive seas around each pocket of land. Put together, the area amounts to over 490,000 square miles of protected ocean (equivalent to three Californias). Industrial

fishing, dumping, mining, and oil and gas drilling were initially prohibited in the reserve, but the degree of protection could change with
a different administration. The area encompasses rare and unspoiled
habitats: underwater mountains (called sea mounts) with colonies
of deepwater corals thousands of years old serving as vital feeding
grounds for migratory and threatened wildlife, including five species
of turtles, silky sharks, beaked whales, giant clams, and manta rays.

The Pacific Remote Islands are a haven for millions of birds as
well, providing undisturbed places for nesting, breeding, or resting
on long migrations. "Jarvis Island alone has nearly three million nesting pairs of sooty terns, which forage more than three hundred miles
from shore even when rearing chicks on the island" (from President
Obama's 2014 proclamation on the Pacific Remote Islands Marine National Monument Expansion).[7]

In global terms the United Kingdom's land area is small, but its
ocean area is prodigious. British overseas territories are primarily
islands, some of which are very remote and encircled by large areas
of ocean, making its total exclusive economic zone the fifth largest
in the world. The government's Blue Belt Programme aims to protect
over 1,500,000 square miles of the marine environment, beginning
with waters around islands and archipelagoes in the Indian Ocean,
South Pacific, and South Atlantic. Introducing the program, the government declared, "These territories and their waters are home to
globally significant biodiversity, from vast penguin colonies in the
South Atlantic to tropical rainforests in the Caribbean. Some of their
species and habitats are found nowhere else on earth."[8]

In the Ascension Island protected area, for example, no commercial fishing is permitted in the surrounding 171,043 square miles of
ocean, thereby safeguarding green turtles, swordfish, sharks, tuna,
and marlins and many bird species including white terns, storm
petrels, and frigate birds. These far-flung wildernesses teeming
with life are nature's security bonds, and their distance from sizable
human settlements, with minimal disturbance, makes them especially valuable. Scientists argue they must be kept intact for our own
long-term well-being—almost like having a nest egg for the future,

safe in the biodiversity bank. But it doesn't always have to be about what's in it for us. Many people believe we must safeguard such wild places simply for what they are, with their glorious profusion of life. That is reason enough.

For an individual country to plan and establish a marine protected area within its own waters, whatever size it is, can be a fairly straightforward, unilateral course to take. For a group of countries to do so is more difficult. In 2002 the Pelagos Sanctuary became the first internationally agreed marine protected area, created by France, Italy, and Monaco for the safety of marine mammals in the northwestern Mediterranean Sea. Other cross-border, cross-sector agreements followed, such as that between the fifteen participating countries of the Convention for the Protection of the Marine Environment of the North-East Atlantic (OSPAR) establishing a network of six protected areas in the North Atlantic (working in cooperation with the Regional Fisheries Management Organization of the area, the North East Atlantic Fisheries Commission). Another is the South Orkney Islands MPA in the Southern Ocean, established by the Commission for the Conservation of Atlantic Marine Living Resources (CCAMLR) and the UK government.

In 2016, CCAMLR raised the bar to new heights of international cooperation, summed up in the *Washington Post*: "World Leaders Create World's Largest Marine Reserve: A historic deal has been hammered out . . . to create the world's largest marine reserve in the Antarctic Ocean."[9] News of the Ross Sea reserve was broadcast around the world on TV and radio, reported in newspapers and on social media. Twenty-five members of CCAMLR (twenty-four countries and the European Union) made the unanimous decision to protect an expanse of sea reaching out from the coast of Antarctica. The reserve covers over 600,000 square miles, most of it fully protected. These waters support large populations of all kinds of sea life, and commercial fishing is banned in most of the reserve, with strictly controlled fishing allowed in some places. The pact to safeguard such a large chunk of the Southern Ocean was variously described as "landmark," "groundbreaking," and "game changing." The creation of the Ross Sea reserve provides another promising prototype of possibility, namely a cross-border commitment to put the well-being of ecosystems and wild species before commercial demands in the global commons. The major flaw in the agreement, however, is that it expires after thirty-five years. Unfortunately, a couple of years later CCAMLR's plans to create another large protected area in the Antarctic were thwarted by Russia, China, and Norway. The Weddell Sea reserve was to have been a no-take zone for commercial activities, including fishing, in order to safeguard key species such as seals, penguins, and whales.

Skeptics question the value of VLMPAs and suspect that their location is determined by wherever it is least inconvenient for industries like fishing and oil exploration, not by the areas' ecological significance. They also suggest that some may be created as nominal acts of conservation, where poor enforcement or minimal restriction on harmful activities means they achieve little. International agreements tie participating states into conservation obligations, and the scramble to create VLMPAs is a relatively easy way for countries to fulfill their treaty commitments and go to the top of the environmental class. There is far less economic and political resistance to protecting seas around remote islands with small populations than there would be for inshore waters close to towns and cities with long-standing fishing traditions and other commercial uses. The debate will no doubt go on. Size matters, some say, while for others it is quality that counts. This is the clash of the clichés. But one characteristic doesn't have to preclude the other. Why not have size *and* quality? Provided the management prevents damaging activities in the designated area with adequate enforcement, then the bigger the better. If we carried on creating bigger protected areas, they would begin to join up like those first raindrops landing on a trail after a dry spell, eventually meeting to become one continuous marine protected area.

It wasn't Einstein who defined insanity as "doing the same thing over and over and expecting different results," but whoever did could have been thinking about how humans use the sea. Beyond the safety of the protected areas, in the rest of the ocean, we know that excessive fishing and pollution have been killing the proverbial goose, yet the "business as usual" approach has continued for so long. Despite this exasperating truth, there are places where people realize that to get a different outcome we have to be smarter and do things differently. More rational approaches to managing how we use the sea are gradually displacing the "in-same" way of doing things. At the international level there is progress with the move toward modernizing the Law of the Sea, as I described earlier. Countries are joining forces, signing treaties, and planning strategies to protect more of the ocean. The Convention for the Protection of the Marine Environment of the North-East Atlantic, the Benguela Current Commission,[10] and the Ross Sea Marine Reserve are three examples. National and regional strategies are increasingly common. They include establishing protected areas, introducing regulations to reduce plastic waste, and putting an end to rapacious commercial fishing. Along with Norway and the United States, other countries are improving fisheries management, among them Belize, Iceland, Ecuador, New Zealand, Chile, Barbados, Senegal, and Namibia. These are all major steps forward, but progress is patchy and slow. National conservation programs need to expand across the board, and they need to speed up.

Alongside improved government policy, there is another, quite different way that positive change comes about for the sea, and this path interested me more. It isn't a command and control, top-down response. It doesn't begin with scientists and civil servants discussing reports and policies in government offices and conference rooms. This is a cumulative drift of change, emerging as people simply get together in coastal towns and villages and decide what to do. The machinery of state can be a dim-witted and cumbersome beast, slow to realize what's going on and impervious to the urgency of a crisis. Citizens and communities want action. They don't want to wait for the beast to get moving and do something after it's too late. When people are let down by the state, they find other ways to solve problems. It may be a group of citizens or even one driven individual who galvanizes a community to take the initiative. People in neighboring areas are motivated to do the same, success stories gather momentum, and they grow into a social movement that spreads like milk spilled on a polished floor.

Ripples Becoming Waves

Community-based marine conservation, also called locally led marine management, is a citizen-driven, hands-on way to safeguard fisheries and marine habitats that emerged in Fiji in the 1990s. The Fijian archipelago of more than three hundred islands is everything you'd imagine as a setting for a Pacific idyll, with coconut palms swaying behind sun-baked beaches and soft coral reefs beneath sparkling seas. As in most small island states, many Fijians are heavily dependent on marine resources for their food and making a living, but by the early 1990s, despite the postcard scenery, it was becoming clear that all was not well in this ocean paradise. People in the village of Ucunivanua on Fiji's largest island, Viti Levu, were finding fewer and fewer kaikoso clams on the shore's mudflats. The clams are a staple food for villagers, and selling the excess is an important source of income. Older residents recalled that in years gone by one person could fill several bags with clams in a day. Eventually it took a whole day to fill just half a bag, and not only were there fewer clams, they were smaller than before. The dwindling of kaikoso clams reflected the populations of other marine species in the area.

Years of unrestrained commercial fishing, of pollution, and of removing live corals and tropical fish for the aquarium market had left much of Fiji's seas and their wildlife in decline. In 1997, with the advice and support of the University of the South Pacific in Fiji's capital, Suva, the community decided to close off sixty acres of nearby mudflats and sea-grass beds for three years. The villagers hoped the closure would allow the clam population to rebound and that larvae would spread to seed and repopulate areas beyond. Twenty men and women made up the area management team. The university made it a condition of its help that men, women, and young people be involved in the project. This set an unusual precedent in Fijian culture, which is traditionally dominated by elder males. It was to be Fiji's first locally managed marine area (LMMA), and the difference it made was dramatic. Within seven years clam populations increased twenty-four times over.[11] Not only were there more clams to eat, the village prospered with higher incomes from selling the surplus. Popula-

Story of how patience pays off

$ ends up not being a problem

tions of other species rebounded as well. The notion that "experts"—scientists, managers, and politicians—know best and should be left to solve problems was debunked. The people of Ucunivanua had made a refreshing departure from the bureaucratic, one-size-fits-all conservation method. As Steve Rocliffe of Blue Ventures reports, "LMMAs put people at the centre: It's the fishers themselves who are making the management decisions, based on their needs, their priorities, and their traditional ecological knowledge."[12]

News of the outcome of the clam experiment reached people in other villages across Fiji that had similar declines in their inshore fisheries and prompted them to follow. They approached the University of the South Pacific for assistance in setting up their own LMMAs. Fiji experienced a quiet, rippling revolution, and hundreds of closed areas have since been set up by communities all around the Fijian coasts.

Locally led marine management inspires and motivates people. They see how quickly depleted populations of marine species can come back, they have the satisfaction of taking control and deciding how to use natural resources in a way that suits them best, and they see the benefits of cooperative decision making and management. "Household incomes have risen up to 30 percent, fish catches have increased, people's adaptive capacity has improved along with their knowledge and attitudes, and communities have regained a sense of ocean stewardship, ownership, and pride."[13]

But was this a rippling revolution or a revival of past practices? For generations village chiefs would order temporary or seasonal closures of important feeding and spawning zones, ban certain fishing practices, and limit the amount of fish and shellfish caught by designating tabu areas. The tabu system made sure that populations of marine species could replenish themselves. The relatively recent development of locally managed marine areas seems to be a return to ancient ways in a twenty-first-century guise. It combines long-established traditions of inshore resource management with modern methods such as scientific assessment and monitoring and wider participation of the community to include women and young people in planning and implementation. Success has been breeding success, and the locally managed marine conservation model has moved well beyond Fijian shores. There are now thousands of similar community-based projects running in island nations and coastal communities all over the South Pacific and far beyond. The movement has reached Papua New Guinea, Indonesia, the Philippines, and right across the Southern Hemisphere from India to Kenya, Mozambique, and Madagascar, as far as South America, Central America through to Mexico and Canada, and from North Africa to Europe.

Grassroots marine conservation is a very positive development in the battle to rescue and conserve the oceans. It has also been shown to reduce poverty, strengthen community fellowship, and raise morale. Government engagement is important too. Coming from opposite ends of society's spectrum, both approaches have their strengths and weaknesses. One lacks legislative weight and sufficient finance, and the other is generally bureaucratic and detached, often based on a planning process that excludes the people whose lives are directly affected. In Fiji the two approaches met in the middle and pulled together, each learning something from the other. The Fijian government now works with fishing communities to further the locally managed marine conservation movement, and the aim is for 30 percent of Fijian seas to be protected in inshore LMMAs. The government was wise enough to recognize a good thing and nurture the seeds of citizen resourcefulness with state support. It also plans to create a network of offshore multiple-use marine protected areas.

To read and hear good news about the sea is a welcome change, because too often bad news runs the show. You see a headline in the newspaper: "Death Blow: Corals and Algae Don't Acclimatize to More Acidic Seas" or "Man Makes Deepest Ocean Dive Ever, Only to Find Plastic Rubbish on the Seafloor." Later there's a program on TV: "Deep Sea Mining: The Latest Threat to Our Oceans," but you turn it off and take the dog out instead. There's a never-ending onslaught of gloomy news about the state of the natural world and the awful things we do to it. Sometimes it's overwhelming; you feel as though we're on a runaway freight train, rushing toward a global catastrophe, and that nothing you do will stop it. Documentaries on climate change, reports on plastic pollution, and high-profile campaigns about the suffering and loss of wildlife have made millions of people aware of the crisis at sea. Understanding is growing, and attitudes about the ocean are changing for the better. We know more about the problems, and to a greater or lesser degree we're recognizing that we're all responsible for them. The dilemma we face is that most people hate to see the natural world emptied of life and choked by waste, yet we're doing the wrecking ourselves simply by living in the modern world. Governments have been asleep at the wheel on major environmental problems, and many people wonder how bad it will have to be before they take eleventh-hour action. Meanwhile, we escape from worry and depressing thoughts by going on vacation, buying stuff, eating stuff, watching stuff, buying more stuff, or just going to work and forgetting about it. Bad news comes to us constantly, like a nagging little demon on our shoulder, bringing us down, sapping our optimism, and lowering our spirits. It can suck out the energy needed to make change happen. Despondency creates a culture of resignation, crushing the potential to come together and build the future most of us want for ourselves and our families. We may grow apathetic and compliant, which suits the establishment very well. Lately there are signs of greater government engagement in environmental concerns, which is encouraging, but genuine preventive and restorative action has yet to be seen. So while the powers that be dither and delay in protecting the sea, what can the rest of us do?

9: The Power of
Many Small Changes

The best thing we could all do to see oceans pollution-free and bursting with life would be to decamp to another planet for a century and give the sea a break. But since that's not going to happen, we need to take the pressure off Mother Nature in other ways so that oceans can thrive long after our children's grandchildren have lived and died. High levels of carbon dioxide emitted by burning fossil fuels accelerate ocean acidification and lead to warmer seas, melting ice caps, rising sea levels, coral bleaching, and life-threatening consequences for people in coastal communities. And as the world's population swells, global demand for fish, oil, and minerals grows, while increasing amounts of shipping and pollution take their toll on the sea. With these factors in mind (although they may seem unrelated to the sea), the most useful steps people can take toward keeping oceans healthy are to slash their use of fossil fuels (and turn to renewable energy sources), eat little or no meat and dairy products, and have fewer babies, because the more of us there are, the more demands we place on already overstretched ocean resources.

Scientists Dirk Notz and Julienne Stroeve, writing in the journal *Science*, found that the carbon dioxide released into the atmosphere on a round-trip flight from New York City to London causes the loss of thirty-two square feet of Arctic ice per passenger.[1] (World Bank analysis shows that the average American is responsible for over seventeen tons of CO_2 emissions each year.)[2] Minimizing air travel means rethinking holidays and vacations and exploring interesting and beautiful places much closer to home. Virtual conference technology makes many business trips unnecessary. Even the Dutch airline KLM is urging people to consider the environmental costs of air travel and to fly less. Cruises are immensely popular, but if you love the ocean, don't go. The ships cause water, air, and noise pollution, frequently damage undersea habitats, and burn large amounts of fossil fuels. Some shipping lines are responding to criticism by incorporating greener technologies into newly built ships. The Norwegian cruise line Hurtigruten is the first operator to power its ships partially with a liquefied biogas produced from dead fish and other organic waste.

The renewable energy industry is growing, but not fast enough. Inexplicably, financial institutions and many governments still subsidize the production of energy from fossil fuels while not investing enough in the development and wider use of renewable energies. The world's oceans will be healthier and have a far brighter future when we release ourselves from an economy based on fossil fuels and move to one driven by clean energy. And is it kismet that the untapped power of the oceans themselves can speed up that transition? They can help us to help them. At present, wind, solar, and hydroelectric power provide the bulk of renewable energy used, but ocean waves, underwater currents, and tides hold immense amounts of energy that is clean, lasts forever, and produces little or no waste. In the quest to find the most efficient way to capture a little of this unending source of power, some intriguing methods have been thought up, sketched out, produced as models in the lab, or built as full-size prototypes for trials at sea, and a great variety of devices have been created, with varying efficiency. They bob, they oscillate, they sway, flutter, and rotate. Possibly the greatest difficulty is to design a device that can withstand the relentless battering from winds and waves in the open sea. These devices are set in tidal barrages, in man-made lagoons, and on the seabed; some float on the surface and others waft in the water column. Energy experts estimate that wave power alone could supply at least twice the world's present demand for electricity.[3] Countries leading the field in developing wave power include the United Kingdom, Germany, Sweden, South Korea, and Australia. China also has ambitious plans to speed up the development of marine renewable energy and is investing billions of yuan into the wave and tidal energy industries. Newer technologies are gaining momentum too. Ocean thermal energy conversion harnesses energy from temperature differences between cooler deep water and the warmer surface waters.

The energy source of OTEC is free, available abundantly and is continually being replenished as long as the sun shines and the natural ocean currents exist. Various renowned parties estimate the amount of energy that can be practically harvested to be in the order of three to five terawatts (1 terawatt is equal to one million million watts). That's almost twice the global electricity demand.[4]

All the world's energy needs can be met entirely with renewable energies, which is very good news, but it won't happen while geriatric coal and oil industries are propped up with public money. Banking adds to the problem. Many banks, pension funds, and insurance companies continue to support environmentally damaging businesses by underwriting loans or with direct investment in operations such as deforestation and fossil fuel extraction. For personal banking it's best to use an ethical bank, to prevent your savings from aiding climate-changing industries. Investors and shareholders with a conscience can put their money where their heart is by using only banks or other financial institutions that invest in environmentally sound companies. An analysis for the *Guardian* newspaper, headlined "Top Investment Banks Provide Billions to Expand Fossil Fuel Industry," revealed the stupefying fact that (at that time) since the Paris climate agreement promising to cut carbon dioxide emissions, banks have provided US$700 billion to shore up the fossil fuel industry. According to the report, Wall Street giant JPMorgan Chase leads the field by supplying US$75 billion to support operations such as fracking and Arctic oil and gas exploration.[5] In October 2019, speaking to a meeting of the House of Commons Treasury committee, Mark Carney, governor of the Bank of England, warned that "the multitrillion-dollar international capital markets—where companies raise funds by selling shares and bonds to investors—are financing activities that would lift global temperatures to more than 4°C above pre-industrial levels," adding that these investments "will keep the world on a trajectory consistent with catastrophic global heating."[6]

The meat and dairy industries put massive stresses on the environment and exhaust natural resources. Research shows that cutting these products from one's diet can reduce one's food-derived carbon footprint by over 70 percent. Meat and dairy production uses millions of gallons of freshwater for irrigation, pollutes river systems and seas with fertilizer runoff, and takes up over 80 percent of the world's farmland, either for grazing or for growing animal feed.[7] It also sends very large quantities of greenhouse gases into the atmosphere, and a growing number of specialists agree that if we all moved to a plant-based diet the planet's health would be transformed—as well as our own.

The third major strategy for easing pressure on the oceans is to slow down population growth (or even better, stop it altogether). The best way to do that is by achieving equal status for women. Birthrates are high in places where women's status is low—countries where girls are unlikely to attend school, child marriage is common, and women have far fewer rights than men. Fewer rights means having less choice about the way you want to live. Greater gender equality enables women to take control of their lives and make choices about their education, their careers, their fertility and contraception—and specifically about how many children to have. It is possible that much wider access to the contraceptive pill could do as much for environmental protection as a whole bundle of international treaties. Families in societies where women have rights and status similar to men's (if not entirely the same) are generally smaller and healthier, with a higher quality of life. In April 2017 the headline of Ian Johnston's article in the British newspaper the *Independent* read, "Women Must Have Equality with Men to Save the Planet, Experts Say." It goes on to explain, "Wherever women are empowered educationally, culturally, economically, politically and legally, fertility rates fall. Giving women equal rights with men is necessary to save the planet from the ongoing mass extinction of wildlife caused largely by the need to grow vast amounts of food for the spiraling human population."[8]

Fewer babies being born results in fewer demands on squeezed resources, and there is another benefit for the global community when women have the same rights and opportunities as men. When women participate in decision making at any level of public or private management, evidence shows that they bring a more cooperative and constructive style of working. Generally they are better communicators and team players and are altogether less commanding than men. History shows that although it is not true of all women, they tend to be gentler and more compassionate leaders with a more inclusive negotiating style. Their influence can moderate the traditional male-dominated, competitive approach. As society's primary nurturers, women at the table also help to ensure that funds and resources are distributed more fairly throughout the community, and the benefits are more likely to be shared.

I thought about how and why society takes the steps that make the world a better place. Change has happened abruptly, following civil unrest and revolution, and it has happened at a plodding pace with a series of small advances over time. Sometimes it has been sparked by an influential book such as Harriet Beecher Stowe's antislavery novel *Uncle Tom's Cabin*, and sometimes it has come after a lodestar court case such as *Brown v. Board of Education*. In 1951 thirteen parents filed a lawsuit against the board of education of Topeka, Kansas, on behalf of twenty pupils. The plaintiffs demanded an end to the policy of racially segregated schools. They won the case when the Supreme Court unanimously ruled the board's policy unconstitutional, signaling the beginning of the end of segregation in the United States. Positive changes often come as a result of education and of sharing information. A tenacious few may expose an aspect of society that is unjust through newspaper articles and advertising, or by speaking on radio and television, or through social media and the internet. A public debate begins, and if the change they are seeking makes good sense, in time understanding and empathy permeate society. People become supportive, and there is a broad shift in attitudes and behavior. After all, most people are fairly reasonable and prefer to live in a society built on a foundation of wisdom and tolerance. Government can be slow to respond, but usually it gets there in the end and introduces legislation to endorse and legitimate the change (which is fine if new laws are implemented and enforced).

In local and national elections, voting for the candidate committed to addressing the causes of climate change and biodiversity loss will help, although something louder, something that becomes a thorn in the authorities' side, is likely to achieve more. Mounting numbers of people, frustrated that authorities don't do more to protect the natural world, are joining forces and becoming very vocal and very proactive. They are forming groups and organizing marches and protests to put pressure on governments. The best known and fastest growing is Extinction Rebellion, which galvanizes thousands of frustrated and indignant citizens into calling for urgent action. They fill the streets chanting, holding up banners and placards, blocking roads, closing bridges, and bringing disruption, spectacle, and entertainment to cities around the world. Most important, they raise the issue so loudly and clearly that it's impossible for those in power not to hear. The real strength of Extinction Rebellion is in the remarkable ordinariness of its supporters, many of them prepared to get arrested during a protest. At any given meeting, or fund-raising event, or protest on the street (which are strictly nonviolent) there will be individuals from all walks of life, of all ages, colors, and religions.

"This is a rebellion / our house is on fire. / If we don't act now / it will be our funeral pyre," a group of over-sixties sang, to the tune of Elvis Presley's "Hound Dog," at the gates of Buckingham Palace in London. One of the group, a retired professor of medicine, said he was there so he could "look his grandchildren in the eye."[9] Challenging a neglectful establishment with a program of direct action transcends the usual societal divisions of class, age, race, and level of income and education, bringing people together in a most unexpected way.

A good way to be a part of constructive change is to support a marine conservation organization by giving a regular donation, keeping informed about its campaigns, and drawing more support by telling other people what it does. Marine conservation NGOs play an important role in pressing legislators to do more for oceans, and they act as a communications bridge between the public and the government. With their scientific and technical expertise, NGOs help shape policies at both a national and an international level (for instance, at the negotiations for a new BBNJ oceans treaty). Besides the international ones like Greenpeace, it's worthwhile to support smaller, locally active organizations and charities, since they tend to be the most financially stretched and, relative to their resources, can be more successful than much larger organizations with high administrative costs. There are many small ocean conservation charities and organizations to support (such as Marinet in the United Kingdom, which I've worked with for many years). One might be active in your neighborhood, or a friend might suggest one. Otherwise an internet search will turn up plenty of options. For example, one of the smaller organizations I'm familiar with is the Olive Ridley Project (https://oliveridleyproject.org/).

When marine biologists Martin Stelfox and David Balson were diving in the Maldives, they were alarmed to find large numbers of olive ridley sea turtles tangled in discarded fishing nets that were carried on the ocean currents. They realized the nets had probably drifted great distances, since in the Maldives fishing is done mainly with pole and line. Martin and David decided to set up the Olive Ridley Project with the aim of diminishing ghost fishing in the Indian Ocean. One of the challenges was to determine where these nets were coming from and how to prevent them from entering the ocean in the first place. By engaging with people in coastal communities, persuading them to adjust their fishing methods and to take part in schemes to recycle and reuse nets, they have been finding ways to reduce the volume of discarded gear getting into the sea. The project also aims to rescue entangled and injured turtles and nurse them back to health. From 2013 to June 2018, the Olive Ridley Project took more than 1,400 ghost nets from the sea and recorded 777 entangled turtles.

As well as (or instead of) donating money to an organization, supporters can get directly involved and volunteer. Conservation NGOs are often underfunded, and most welcome regular volunteers to help with answering emails and phone calls, updating websites, writing press releases, planning fund-raising events, or perhaps helping a team collect plastic debris along the shore. Joining a scientific expedition and being active on the front line is another way to get involved. Several organizations run projects, among them Coral Cay Conservation Expeditions (https://www.coralcay.org/), Marine Conservation Philippines (https://www.marineconservationphilippines.org/), and Blue Ventures (https://blueventures.org/).

Beach Recovery

Most people don't care to see the shoreline littered with waste, plastic or otherwise, so group cleanups have gone global. They are set in motion by local residents, conservation charities, and sometimes district authorities. The procedure doesn't vary much: it's basically a bunch of people getting together on a particular day or a series of days and working en masse to clear beaches that are under trash attack. The International Coastal Cleanup is part of a campaign run by the American charity Ocean Conservancy. "Trash Free Seas" aims to get citizens engaged in hands-on conservation across the world by organizing events to rid the shores of rubbish and record what is collected.[10] Other organizations set up similar cleanups in their regions, such as Legambiente in the Mediterranean, Surfriders in the United States, Clean C in South Africa, the Marine Conservation Society in the United Kingdom, and One Island One Voice in Bali. Volunteers make a detailed record of the trash to give a clearer picture of where the items come from, helping to understand the problem and how to tackle it. Among the most commonly collected items are cigarette butts, plastic drink bottles and caps, food wrappers, plastic grocery bags, and drinking straws. The International Coastal Cleanup also has a survey category called weird finds, which has included entries like refrigerators, chandeliers, and even a grand piano.

Apart from the International Coastal Cleanup, one of the most optimistic cleanup stories comes from India. Young lawyer Afroz Shah remembers that as a child growing up in Mumbai, on India's west coast, he swam in clear waters and played on a clean beach. After some years away, he returned to the city and was appalled to see that Mumbai's Versova Beach was like a garbage dump, its white sands all but covered with plastic rubbish, mostly washed up by the tide twice a day. Afroz decided he had to do something about it. In 2015 he began knocking on doors and badgering other residents to help him clean the beach. He rallied the support of hundreds of people, and together they cleared 4,400 tons of waste in a year, most of it some sort of plastic. It has been called the world's biggest community beach cleanup. These days the shore is almost pristine, and children can paddle in the water and play in the sand again. Cleaning has to continue, though, as the sea constantly brings in more debris. For his tireless efforts, in 2016 Afroz was made one of the United Nations' Champions of the Earth. "Single use plastic has no place in the 21st century, no place in this world," he says, "the ocean is home for fish, marine creatures and birds. It's where they belong, they live, they survive. I don't take it kindly. No species has the right to destroy another's home."[11]

In 2018, less than three years after the beach cleanup began, olive ridley turtles returned to Versova Beach. About eighty hatchlings were seen emerging from their sandy nests and scrambling toward the sea. It was the first time baby turtles had been seen on the beach in twenty years.

Picking up crumpled coffee-cup lids and plastic bottles, old lighters, and unloved toys out of the sand and carrying away sacks of debris is strangely confusing, bringing up conflicting emotions. You feel embarrassed for your species, guilty. You feel like part of a great tribe of wanton yobs darkening these beautiful places with waste. But it feels good to be doing something useful and straightforward that has immediate results. As you pick up discarded bottles and pull up half-buried plastic bags, you look along the beach and see the rest of the team doing the same thing. The beach cleanup movement mirrors the big-picture solution. It sums up how to solve all the threats facing coasts and oceans: by thinking big, having a forward-looking plan, and working together.

Being conscious of our daily impact on the ocean and making modifications to one's life's routines is an easy way to move from compounding the ocean's troubles toward alleviating them. In many cases it's easy to see the connection between what we do and what happens in the sea. Buying souvenirs like jewelry made of coral and dried seahorses, or using health and beauty products derived from heavily depleted species such as shark cartilage and shark liver oil, is obviously not the thing to do. Recreational sea fishing is hugely popular, with millions of people taking part around the world. Anyone who fishes should be aware of the local and regional regulations and should follow them (which might include buying a license to fish). It's best to fish only for species that are abundant, and to be sure not to leave behind any tackle that will endanger wildlife or people (lines, hooks, weights, sinkers, etc.). Fishers who are catching and releasing should use circle or barbless hooks and handle the fish with great care, releasing it as soon as possible and preferably not removing it from the water. The American government agency the National Oceanic and Atmospheric Administration (NOAA) has an excellent page of catch and release best practices on its website.[12]

As for big-game fishing for the likes of marlins, sailfish, swordfish, sharks, and large tuna species, with 90 percent of the ocean's top predators fished out, can't we leave them alone? Even if they're released after capture, many fish don't survive, particularly if there has been a prolonged struggle to catch them. For an alternative seagoing adrenaline fix, would-be game fishers could turn their energies away from an activity that reduces big fish numbers to one that helps their populations recover. Project Aware (https://www.projectaware .org), for instance, has the tagline "Where Conservation Meets Adventure" and engages scuba divers in "citizen action in science" to protect undersea habitats, with a particular focus on shark conservation and ocean litter. Volunteers help to gather data on underwater habitats and species, clean reefs, collect marine debris, and raise public awareness.

Like fishers, those taking boats and yachts out to sea also have a responsibility to be mindful of the marine environment and wildlife. Codes of good practice for boaters include reducing the risk of oil leaks and of transporting invasive nonnative species,[13] keeping a safe distance from wildlife, recycling monofilament fishing line, and not discarding litter overboard.

Being better informed means we can be wiser shoppers. For people who eat fish, knowing which type to buy is useful, whether you eat in a restaurant or cook for yourself. Making the right choice as a consumer helps to conserve fish populations, protect habitats, and prevent illegal fishing. Take tuna, a fish enjoyed around the world, canned, fresh, or frozen. Figures from the International Seafood Sustainability Foundation (ISSF) tell us that 5.3 million tons of tuna were landed in 2017 (all five commercial species combined).[14] But if you eat tuna, which kind can you buy with a clear conscience? Certainly the

American Albacore Fishing Association's certified albacore, and probably Maldives pole-and-line-caught tuna; then it gets complicated and information is conflicting. Tuna caught with pole and line are the most environmentally sound because they are taken from the water one by one, whereas anything caught in a net will take a collateral toll on nontarget animals. Approximately 11 percent of the world's tuna catch is caught on longlines and 65 percent of it in purse seines: huge circular nets set around man-made rafts where fish and other animals congregate (called fish aggregate devices, or FADs).[15] The purse seine encircles a whole school of fish and is then drawn tight at top and bottom like an old-fashioned purse. As Andrew Balmford reports, "Purse-seining around FADs catches lots of tuna and very few dolphins, but even more by-catch of other animals (turtles, sharks, swordfish, marlin, barracuda, manta rays . . .). Dolphin friendly maybe, but distinctly unfriendly to just about everything else in the sea."[16]

Besides that, when an entire school is taken, who is left to reproduce? The ISSF found that only 8 percent of tuna were caught with a pole and line. So when it comes to deciding which tuna to buy, I find it easier just to eat something else.

The Marine Stewardship Council (MSC) assesses standards of fisheries management around the globe, and a fishery must pass the accreditation process to be able to add the MSC's blue fish sticker to its products, assuring consumers that it was responsibly fished. The process inspects three aspects of a fishery's sustainability: its impact on the target stock, how well the fishery is managed, and how it affects the marine habitat and ecosystems. Since a growing number of supermarkets and restaurants are buying fish only from sustainable sources, the blue sticker makes certified fish more marketable and more valuable. The MSC has been rebuked by some conservationists for being "captured by industry," questioning some of its regulations and a few surprising accreditations. One such was for the orange roughy in New Zealand, an extremely slow-growing species that is especially vulnerable to commercial fishing and caught by bottom trawling, which has been likened to bulldozing the seafloor. In 2018, in an open letter to the MSC, signed by representatives of over sixty conservation bodies, the organization was taken to task for certifying a number of fisheries as being sustainable in spite of the damage caused. Among the criticisms were that the controversial fisheries catch thousands of vulnerable and endangered animals, routinely discard and waste considerable amounts of sea life as unwanted bycatch, and destroy vulnerable seabed habitats. The letter calls for "urgent implementation of critical improvements" to the MSC's certification policies. Some experts argue that the MSC should not certify any industrial-scale fishery or one that uses any form of trawling. The Marine Stewardship Council listened to its critics and issued a public statement on plans to improve the certification process. The World Wildlife Fund's response to the MSC's "we will do better" statement has the air of a teacher reprimanding a wayward pupil. "While WWF welcomes the MSC's commitment to action various improvements, there are key areas where we believe the MSC must make rapid and clear progress."[17]

The conclusion of all this is that an MSC blue logo on a product may not be a cast-iron guarantee that that particular fish has come from a responsibly managed fishery, but it's much more probable than if there is no sticker. I would always avoid the orange roughy, though, whatever sticker it has.

When deciding which fish to buy it makes sense to investigate for yourself, because populations of the various species fluctuate. The sustainable option depends on where you live and the latest stock data for each species. Seafood advisory sites like Monterey Bay Aquarium's Seafood Watch in the United States and the Marine Conservation Society's Good Fish Guide in the United Kingdom make recommendations for individual customers, chefs, restaurants, and businesses on which fish to buy and which to steer clear of. For people living near the coast, locally caught inshore fish is usually a good option. Shifting tastes away from the bigger species and learning to enjoy small, abundant ones like herring, sardines, mackerel, and anchovies is also helpful. Bearing in mind that thousands of tons of anchovies are made into fish meal for carnivorous farmed fish like salmon and sea bass, Oceana's CEO Andy Sharpless suggests cutting out the middleman (or middle fish) altogether. "Just get people to eat anchovies instead of feeding anchovies to salmon."[18]

Mangrove Forests

Coral reefs and tropical forests are well known for their wealth and diversity of life. A less celebrated but invaluable habitat is mangrove forest. Although this is one of Earth's most productive and complex ecosystems, many of us are unknowingly playing a part in its widespread decline. Especially adapted to live in the intertidal world of salt and water, between land and sea, mangrove trees and shrubs grow on sheltered shorelines, in lagoons, and along the edges of estuaries in tropical and subtropical regions around the globe. They cluster into a dense belt of coastal woodland. At low tide a chaotic jumble of leggy roots is exposed, arching high from the muddy bank up to the leaf canopy above. Few land plants can cope with salt water as frequently as with the twice-daily invasion of incoming tides, but mangroves filter out the salt and expel it through pores on their leaves. The trees don't just cope with the salty conditions, they thrive on them. An extraordinary range of creatures live in mangrove forests, including all kinds of fish, amphibians, reptiles, insects, birds, and mammals. Under the water's surface, tangles of roots provide vital spawning grounds for a great many species of fish and crustaceans and give sanctuary to juvenile fish until they are mature enough to survive in the open sea. Marine biologists estimate that about three-quarters of tropical fish species begin life in these brackish waters. The underwater roots are encrusted with barnacles, oysters, mussels, sponges, anemones, and many other organisms. Crabs love mangroves; among them are grapsid and fiddler crabs, which live in dense populations on the water's edge.

The root systems filter toxins from the water and stabilize soft sediments in tidal beds that would otherwise be eroded away by the constant movement of the waves. In this way they also prevent nearby sea-grass beds and coral reefs from being smothered by waterborne sediments. Mangrove forests create a defensive barrier (particularly beneficial in low-lying areas) that heads off flooding and protects coasts from severe weather: incoming storms, hurricanes, and tsunamis. Around the world, the forests' wild fish and shellfish populations have also provided generations of fishers and their families with food and income. And there's more: mangroves are very good at absorbing carbon dioxide and storing it. "Blue carbon" is carbon dioxide captured from the atmosphere and stored in oceans and coastal ecosystems, in the sediments and biomass of habitats such as mangrove forests, giant kelp forests, sea-grass beds, and intertidal salt marshes. Mangrove forests store up to four times as much carbon as land-based forests, putting them on the front line in combating climate change. Most such forests form a strip along the coast, typically less than two miles wide, but some are very extensive. Straddling the border of India and Bangladesh, where the Ganges, Brahmaputra, and Meghna Rivers converge, the Sundarbans is the world's largest mangrove forest. Covering 3,900 square miles, it is a vast network of tidal waterways, mudflats, and small islands that is home to a great variety of wildlife, including 260 bird species. Some of the Sundarbans residents are rare: the estuarine (or saltwater) crocodile and Ganges River dolphins can be found there, and the iconic royal Bengal tiger.

In ecological ranking the fifty-plus species of mangroves are superstars. But their tremendous value was late to be recognized, and approximately half the world's mangrove forests have been cleared over the past half century for timber, for coastal development (buildings, roads, marinas, golf courses, etc.), and mainly to make space for farming shrimps and prawns—mostly sold in Europe, North America, and Japan. Thousands of acres of mangroves are lost each year to make way for aquaculture ponds, particularly in China, Thailand, Indonesia, India, and Vietnam, and with some farms in Latin America. Pesticides, fertilizers, and antibiotics are added to the water, and after three years or so the ponds become too toxic and are abandoned. More mangroves are then removed, and new shrimp and prawn aquaculture ponds take their place. And so it goes on. Since commercial farming took off in the 1970s, global production has risen sharply, driven by the growing demand, making it the leading cause of mangrove clearance. And while coastal communities are losing the forests that helped sustain and protect them, all too often large profits from the shrimp and prawn farm industry fill the pockets of distant investors.

Initiatives such as the Mangrove Action Project and the Global

Mangrove Alliance (a partner of the WWF), Conservation International, the International Union for the Conservation of Nature, and the Nature Conservancy are working to reverse the loss of mangroves. Action to rescue mangroves ranges from ambitious projects by large NGO alliances with a transborder reach to hands-on teamwork at a local level, similar to the clam fishers' project in Fiji. A good example is the Society for Women and Vulnerable Group Empowerment (SWOVUGE), which is "a women-led civil society group, empowering women and the whole community to protect Nigeria's extremely productive but disappearing mangrove forests."[19] Families in the area depend on the natural resources of the Ukpom Community Mangrove in southeastern Nigeria, and the forest also supports a wealth of wildlife species including crocodiles, tortoises, turtles, fish, snails, shrimps, crab, clams, and oysters. But over the years too many trees have been taken down for firewood, to dry fish, to build canoes, or to make space for building. Now SWOVUGE is helping people from five villages to replant mangroves and use forest resources with a view to the future. Some enlightened governments are also rising to the challenge. The Ecuadorian Ministry of Environment, for example, supports community-led restoration projects such as the Isla Corazon Mangrove Reserve, set up by local fishers on an island in the river Chone estuary, which flows into the Pacific. A wide range of bird species, including magnificent frigate birds, brown pelicans, snowy egrets, little blue herons, and white ibis, have made the island a magnet for bird-watchers. In 2015 the Sri Lankan government took things a step further by becoming the first country to protect all of its remaining 21,800 acres of mangrove forests. The scheme will provide 15,000 women with small loans to set up businesses (beekeeping, for instance) in return for conserving remaining forests, and another 9,600 acres of mangroves are due to be replanted.

Apart from supporting a mangrove conservation project, the best way we, as consumers, can help protect these forests is by boycotting warm-water-farmed shrimps and prawns (unless they came from a genuine responsibly managed source). Before ordering that prawn curry or shrimp gumbo in the restaurant, ask where the shellfish came from, and check the packaging of products in the grocery store or supermarket. Shrimps and prawns coming from a tropical or subtropical country without bona fide sustainable certification are probably from a mangrove-cleared fish farm, as the vast majority are. Tell the restaurant or store manager why you want to know and why you will choose only fish that is assured to have been responsibly farmed. In fact, tell everyone you know.

We don't only buy fish to eat, of course, we also buy them to watch. Tropical fish are beautiful and fascinating, darting around illuminated tanks, brightening up living rooms and hospital waiting rooms across the land. But enthusiasts might not realize that their love of undersea life could be contributing to its depletion in the wild. A great number of aquarium fish are captured in their wild habitats, mostly on the reefs of Indonesia and the Philippines and in locations such as Hawaii, Fiji, and Kenya. Often they are caught using cyanide, which is sprayed on the reef to stun the fish, and it frequently kills nontargeted species and corals. It is estimated that half the target fish die either when they are captured or in transit after being bagged and boxed. So it's important to buy only captive-raised fish sold by a reputable aquarium supplier.

There are other, less obvious shopping choices that reduce damage to the sea. Perhaps surprisingly, even our clothes are a threat to ocean life, both before we buy them and after we wear them. The textile dyeing and printing industry uses more than eight thousand chemicals, many of them toxic. The World Bank estimates that 17 to 20 percent of industrial water pollution running into rivers and then into the sea is caused by dyeing and finishing processes. The industry also takes up enormous quantities of water for the various processes: four to six gallons for each pound of textiles produced.[20] A large proportion of garments are made with synthetic polymers containing life-unfriendly substances. Most fleeces, for instance, contain polyethylene terephthalate (PET), the same substance used to make plastic bottles. In being worn or washed, these fabrics shed thousands of tiny plastic fibers, volumes of which end up in the sea. If washing machine manufacturers fitted appropriate filters by default, the amount of plastic

microfiber entering natural water systems would be dramatically reduced. The textile and wastewater industries could also cooperate to
introduce more rigorous filtration and decontaminating systems to
prevent toxins and microfibers from entering rivers and seas from
the point of production. Being a victim of fashion, constantly buying
new clothes and wearing them just a few times, further undermines
ocean health. It is better to buy good-quality clothes made with sustainably produced natural fabrics in natural colors and wear them for
the duration or—even better—buy secondhand.

Stores of Poison

Like many of us, I live in a culture that is nearly obsessed with keeping
everything clean. We're forever washing or whitening, scrubbing, ster-
ilizing, and perfuming ourselves and our belongings. But the cleaner
our houses and clothes and bodies are, the sicker the seas become.
The peddlers of spotlessness fire out endless advertising campaigns,
expecting us to believe that if the kitchen floor isn't gleaming, if our
hair isn't glossy and the bedsheets aren't blindingly white, then our
lives are barely worth living. So certain substances in everyday household products are constantly washed down drains, into rivers, and
eventually into the sea, harming or killing marine life.

I walked around my local supermarket and checked out some of
the cleaning products on the shelves, looking at dish and laundry
detergents, fabric softeners, stain removers, toilet cleaners, mold removers, and air fresheners. Most of these listed at least one ingredi-
ent known to poison marine life. A surprising number of packages
and bottles had a macabre symbol on the back label as well—a dead
tree and a dead fish set in a bold red box, with DANGER written in
capital letters below. The instructions direct you to call the Poison
Center immediately if anyone happens to swallow a little of the contents. One item in particular caught my attention because it was frequently advertised on television at the time: a bottle of something
with the unfortunate name of Unstoppables. The manufacturer describes these as "in-wash scent boosters that you pop in your washing

machine drum to add up to twelve weeks of boosted freshness." On the label it states unashamedly, "Harmful to aquatic life with long-lasting effects." I find this stuff especially objectionable not because it's necessarily any more dangerous than the other products, but because it's such an idiotic concoction. Large sums of money are spent on developing, marketing, and promoting a plastic bottle full of poisonous chemicals compressed into little nuggets of nastiness to make your laundry smell synthetic—the implication being that perfumed pants are worth more than vibrant seas.

Then I moved on to the health and beauty section. There were lots of unlikely villains here too, standing patiently on the shelf in tubes and bottles waiting to be released by their unwitting owners and washed away to do bad things in the sea. There were toothpastes, soaps, and gels containing the antibacterial agent triclosan, shampoos and bubble bath with parabens and sodium lauryl sulfate, hair dyes with p-phenylenediamine, deodorants with aluminum, and an array of sunscreens containing that dreadful duo oxybenzone and octinoxate. All these chemicals (and more in other products) impair marine life to some degree. Fortunately, "rinse-off" personal hygiene products (such as facial scrubs, soap, and toothpaste) that contain thousands of tiny plastic microbeads have been banned in some countries because of their lethal effects on sea life. These countries include the Netherlands, Sweden, the United States, Canada, the United Kingdom, and New Zealand. It's also worth considering that whatever is harmful to marine creatures is unlikely to be much good for whoever is putting it on their clothes, their skin, and their hair.

I left the inappropriately named health and beauty section and headed for the exit, passing the cards, stationery, and party accessories aisle on the way. Here there were more ocean vandals. Besides the plastic pens, tiny plastic toys, and neon paints, there were all sorts of products dipped in glitter: greeting cards, balloons, confetti, makeup, face paints, and whole tubes of glitter to scatter wherever you wish. Their appeal to a child makes "all that glitters" an insidious diversion, their allure being inversely proportionate to the harm they inflict if the glitter ends up in the ocean after the party is over.

Most glitter is made from plastic, and when it disappears down the drain toward the sea (as much of it will) it becomes a different form of microplastic marine litter—glitter litter if you like. It can be eaten by plankton, fish, shellfish, seabirds, and other marine life, poisoning them or contributing to their starvation. Biodegradable, ocean-friendly glitter alternatives are available, and it's worth urging suppliers to stock those rather than the plastic varieties, and to make local schools and preschools aware of the issue too.

For the final leg of my household chemical mission, I trekked on to the home improvement store. Here I found more menacing liquids, poisonous powders, and noxious paints. There were all kinds of miscreants used to wash brushes, preserve wood, strip paint, unclog drains, fertilize the grass, and kill the weeds. These liquids, often unintentionally, are likely to end up in the sewer or leached into the soil, from there to run into watercourses, rivers, and finally the sea. The message is, know your chemicals and read the small print. And if it isn't possible to avoid the nasties altogether, we shouldn't let them be washed down the drain. (Local authorities can usually advise on where to dispose of potentially toxic substances safely.)

In short, when thinking about what an individual can do toward conserving oceans, it's worth looking at nearly every aspect of daily life. In the words of Philippe Cousteau, "When considering your impact on the planet, realize that the critical concept is not that you can make a difference; it's that everything you do already does makes a difference."[21]

Many people are happy to assess their lifestyles and day-to-day habits and are prepared to make adjustments once they know that seemingly minor decisions, such as choosing the wrong sunscreen, can poison marine life hundreds of miles away. Others are prepared to do much more. When people hear about positive change, they're inspired to get involved to improve their surroundings and their futures. For some it's the motivating power of hearing good news that gets them off the sofa to join in. For others it's hearing bad news that does it. Award-winning campaigner Natalie Fee told me what pushed her into action.

I watched a trailer for the film *Midway*, by the artist Chris Jordan. It had images of Laysan albatross chicks that were starving with their bellies full of plastic, and they weren't ever going to leave the nest. I wasn't an environmental campaigner at the time, but there was something about seeing these beautiful birds and watching footage of them dying that was heartbreaking. I was grief-stricken watching this human-made environmental disaster unfold. I'd never seen something that moved me on that scale before. Seeing everyday items that I was using—a toothbrush, bottle tops, plastic toys, lighters, ink cartridges—things that I used were killing these majestic seabirds who lived on a remote atoll over two thousand miles from the nearest continent. I just felt, "I can't let that happen. I don't want those birds to be wiped out because of all the plastic we're using and not properly disposing of."

Natalie wanted to raise the profile of the plague of plastic in the sea, because at the time little seemed to be known about it. She decided to set up a nonprofit in Bristol, England, to make people aware of the issue, and she formed City to Sea (https://www.citytosea.org .uk/). With the help of local environmental organizations, scientists, and marine biologists, Natalie organized some public consultations and found that people were most concerned about plastic bottles (and bottle tops) and cotton swabs (cotton buds). Bristol lies on the river Avon, and residents could see its banks littered with empty bottles and thousands of used cotton swabs. When flushed down the toilet, cotton swabs are small enough to slip through sewage filters, and they escape into the natural water systems, collecting on riverbanks and beaches or entering the sea. Natalie then started campaigns focusing on these two issues.

Refill is our free tap water campaign, which is designed to give anyone who is on the go easy access to refill their water bottle. We got some funding for a trial from Bristol's grants scheme to support sustainable projects, and within three months we had about two hundred refill stations across the city—that is, shops, cafés, hotels, and pubs, all displaying a blue sticker in the window and a friendly poster inviting people to come in and refill their bottles.

With financial support from local businesses and the water supply industry (which wants to encourage customers to drink more tap water), the Refill campaign has gone international as other cities and regions have become involved.

At the same time as running Refill we started also putting our energy into Switch the Stick, which was our campaign calling on all UK retailers to switch from selling plastic-stemmed cotton buds to paper-stemmed cotton buds.

Natalie and her small team set up an online petition. A month and six thousand signatures later, the online campaigning organization 38 Degrees approached City to Sea and offered to send the petition to their supporters. Naturally she accepted.

We went from 6,000 signatures to 155,000 signatures in about six weeks, and then all the supermarkets started listening. Tesco, Sainsbury's, and Asda were the first to agree to make the switch to paper-stemmed cotton buds. After that, the rest followed. We worked out that supermarkets' making the switch will reduce production by about two billion plastic sticks a year. So we were really chuffed!

City to Sea is also pressing for a reduction in the amount of menstrual products and wet wipes being flushed away, since these contain significant quantities of plastic. The campaigns are remarkably effective considering the size of the organization and the limited resources available. Plus, things happened quickly. It took less than a year to get all the major supermarkets in the United Kingdom to switch from plastic- to paper-stemmed cotton swabs—quite an achievement for such a small operation. It reveals the potential power of one person's passion and drive, coupled with an understanding of the cultural mood of the times. In 2017 Natalie won the Sheila McKechnie Award for Environmental Justice in commendation of her work, and with success and more funding, the City to Sea team has expanded to a staff of over thirty. In the battle against the ubiquitous plastic bottle, the aim is to take the Refill campaign beyond tap water, so that people can find places to refill their coffee cups, lunch boxes, and even containers of household items like laundry liquid and groceries.

I finished by asking how she would advise someone who felt inspired and wanted to be more proactive. She said,

> Become a reuser; have a reusable bottle and coffee cup, so when you're out and about people see you—that's a big step in normalizing refilling. I would say, support an organization that's campaigning for cleaner seas, lobby the local government to become plastic-free, and try to live for a month without single-use plastic.

The power of many small changes is indisputable. If a million individuals stopped buying those noxious household potions and lotions and drank from a refilled bottle instead of buying a plastic one, if we ate a lot less of everything (particularly meat and dairy products), if we chose to avoid all short-life plastic products and packaging, if we bought only wild fish from well-managed fisheries, if we flew less, drove less, walked or cycled to work and school, and traded in our gas-guzzling SUVs for cars powered by clean energy, if we heated and cooled our homes with renewable energy and made sure they are well insulated, and bought fewer clothes, fewer gadgets, fewer household furnishings—if we sought contentment beyond materialism and bought altogether much less stuff—the accumulated result of all these choices to take the pressure off Mother Ocean would be substantial. If a hundred million individuals did the same, the end result would be momentous. They aren't radical lifestyle changes; mostly they are tweaks to habits. If enough of us stop buying the damaging products, manufacturers will stop producing them and will have to develop benign alternatives. But will enough people make those lifestyle changes voluntarily? It's often simpler to keep on doing what you're accustomed to. Neither is it made easy for people to adjust their present ways and break away from the conventional consumerism that defines society today. There are too few incentives and too many dis-incentives to living in a nature-friendly way. Living free of fossil fuels, plastic, or chemicals is generally more time consuming and more expensive than not doing so. Governments, business, and industry can make it easier for citizens to do the right thing and harder to do the wrong thing. People need to benefit from adapting their lifestyles to safeguard nature and not lose out by doing so.

As I walked back past the store full of poison, I mulled over all these things. As discerning consumers we have some leverage to make positive changes, and it's important to use that leverage by bypassing products that damage the ocean and sea life. Although I agree with that, I can't imagine shoppers' carefully reading a long list of obscure chemicals, usually written in tiny type, on the back of every bottle or package before deciding whether to buy it. Most of us are not chemists, and we don't know what the ingredients are or what effects they have. And few of us have time to stand and scrutinize all those labels. According to international law, manufacturers should not be using chemicals that harm ocean life, and consumers shouldn't be able to buy them in any product form. Besides the Law of the Sea's requirement to prevent marine pollution from land, two international treaties—the London Convention (1972) and the Basel Convention (1989)—were intended to prevent pollution and hazardous waste from entering the sea. In addition, in 2019, 187 countries agreed to include plastic waste in the Basel Convention and to make greater efforts to stop its moving across national borders or into the sea.[22] We only hope it won't be another noble but empty pledge that doesn't deliver genuine change.

Consumer choice can do only so much to kick out ruinous systems or products. There are others in society who have the power to do a great deal more to safeguard the sea. They are company directors, investors, lawyers, shareholders, and board members of large corporations from a range of sectors: energy, mining, construction, transport, fishing, shipping, chemical, fertilizer, and plastics manufacturing, and retail, catering, and tourism. Among those with the most power to change minds are journalists, news editors, and filmmakers. In Britain, broadcaster and naturalist Sir David Attenborough is the most loved and respected man in the land. A single episode of his BBC television series Blue Planet II, aired in autumn 2016, is legendary for the change it made to the national psyche with regard to single-use plastics. A hawksbill turtle was seen tangled up in a plastic sack, an albatross unwittingly feeding plastic to its chick, a sperm whale trying to eat a plastic bucket. Viewers were shocked and ashamed. Since then not a week goes by without a feature in the media on the scourge of plastic waste and what we can do about it. Public pressure on politicians and retailers to act has seen new laws banning the manufacture and sale of plastic drinking straws and cotton swabs, with the promise of more legislation against plastics to come. Supermarkets are reducing plastic packaging for many items or cutting it out altogether. People have altered their buying habits and their attitudes about individual consumers' responsibility. One survey found that 88 percent of those who watched that Blue Planet II episode have adjusted their lifestyles with regard to single-use plastics.[23]

Ultimately, though, real power to make the shift to an economy based on clean energy, to outlaw destructive industrial practices, stem the tide of poisons and waste entering the sea, and manage our use of natural resources wisely lies with government: with civil servants, policy advisers, ministers, senators, members of Congress, the judiciary, and political leaders. The great mystery of our times is why they don't do more right now to prevent overfishing, pollution, and the causes of climate change, plus innumerable other solvable environmental problems—in the sea, on land, and in the air. Governments aren't made up of some alien or supernatural force—they're only groups of people, after all.

10: Finding Like Minds

For a couple of years I had been gathering evidence to support a case for reinventing the ocean's "normal" and safeguarding the whole lot from destructive practices. I had talked to all kinds of people who are related in some way to the sea or to nature conservation, including a few influential figures, to test the waters and generate support. Oliver Tickell, former *Ecologist* editor and seasoned campaigner; Charles Secrett, former CEO of Friends of the Earth UK; George Monbiot, *Guardian* columnist, author, and political activist; Dr. Stephen O. Andersen and Dr. Suely Carvalho, key figures from the Montreal Protocol; and Polly Higgins, *Ecocide* lawyer, have all expressed their enthusiasm and support for the proposal. Meanwhile I've met dismissive indifference with a "don't bother me now" expression from others. Backing from respected journalists and conservation campaigners is invaluable, and authoritative scientific endorsement gives the concept extra weight. But one leading marine biologist I put it to rejected the idea, insisting that the only solution was reforming fisheries management and creating more marine reserves. With another I fared better. He said yes, we should be aiming to use all of the sea in a responsible way, but the idea was too big a leap, and because of his professional position he didn't want his views made public. Ho hum, I thought, disappointed but undeterred.

There are scores of scientists working in all areas of marine conservation. During my research a handful of names cropped up repeatedly; Callum Roberts, Ruth Gates, Boris Worm, Rainer Froese, and Chris Costello, to name a few—as either sole authors or coauthors of books, research papers, and articles in newspapers or journals, and on websites. However, the names I see most frequently are Daniel Pauly and Rashid Sumaila.

Daniel Pauly was born in Paris, raised in Switzerland, studied fisheries science at Kiel University in Germany, and earned a master's degree with his fieldwork in Ghana. Consequently he had an unusually international start in life, and this seems to have been formative for his work. He says, "I wanted to study an applied discipline so that I could work and contribute to a—yet unidentified—developing country, to which I would eventually emigrate."[1] After a spell conducting trawl-fishing surveys in Indonesia (which inspired his doctoral dissertation back in Kiel), he worked for many years at the International Centre for Living Aquatic Resources Management (ICLARM) in the Philippines, developing new ways of estimating fish populations and gauging their growth and mortality. Since 1994 he has been professor of fisheries at the Institute for the Oceans and Fisheries (IOF; formerly called the Fisheries Centre) at the University of British Columbia (UBC) and principal investigator for the Sea Around Us Project. To chart his career and achievements would take up a whole chapter, if not a whole book, but one of Daniel Pauly's most significant contributions is the Sea Around Us Project, a research initiative he founded in 1999, which created an online database of world fish catches. Figures date from 1950 to the present, combining official landings figures with estimates of unreported landings and discards. The data are freely available to all (fishers and fishery managers included); the purpose is to assess "the impact of fisheries on the marine ecosystems of the world, and [offer] mitigating solutions to a range of stakeholders."[2] In his own words,

At UBC, I began to assemble a brilliant group of people (Villy Christensen, Rashid Sumaila, Reg Watson, Deng Palomares, Dirk Zeller, and others) and design a project called the Sea Around Us (named after Rachel Carson's last book) to document the state of the oceans and to work with civil society to help slow down or reverse negative trends. One of our first visible results was an analysis of world catch trends which demonstrated that the world catch was declining, as could be inferred by negative stories everywhere, but that had been masked for years by catch over-reporting by China. This came at about the same time that Jeremy Jackson and his colleagues were showing that overfishing had been the modus operandi for millennia, and also at the same time that papers of the late Ransom Myers were hitting the news. Jointly, these papers changed the view that fisheries are isolated affairs, failing separately from each other. Rather, there is now the realization that our entire mode of interaction with the sea is wrong; just like we don't believe any more that this or that bank failed, but that the whole financial system failed us.[3]

These days he is stepping back a little from university responsibilities to spend time on more personal interests, one being a return to research he began in the late 1970s on the relation of fish growth to the surface area of the gills and to water temperature (a subject that has become more topical since the effects of climate change and global warming are more evident).

Daniel Pauly is a people's scientist, a maverick. Scientists traditionally tend to be apolitical and prefer to sit on the fence, whereas he is openly critical of governments that fail to prevent excessive commercial fishing (and even prolong it with supportive subsidies) and of fishing corporations that ignore scientific advice on catch limits and continue to demolish ocean habitats.

> The start of my work at the University of British Columbia (Fall 1994) coincided with a decision to become more directly involved in conservation issues with environmental nongovernment organizations (NGOs). . . . To protect fisheries from their own suicidal tendencies (and to protect the biodiversity and ecosystems they depend on), outside pressure must be applied. Such pressure is best generated by civil society, i.e., by conservation-oriented or, generally, by environmental NGOs.[4]

No one working in ocean conservation today has more authority or is more respected than Daniel Pauly. What would he have to say about the proposition to protect all of the sea instead of only parts of it? I emailed him outlining the whole-ocean idea and requested a meeting. His personal assistant Valentina replied, suggesting a date a few weeks later. I was happy with that because, fortuitously, I had planned to visit my daughter Ellie, who was working in Vancouver.

It was a sparkling autumn morning when I arrived at the University of British Columbia. The Vancouver campus is like a miniature city, with broad, leafy walkways linking lecture halls, libraries, and coffeehouses, all wedged between the ocean and forest parkland. I found Dr. Pauly's office in the Institute for Oceans and Fisheries easily enough. Leaning tentatively around the partly open door, I knocked gently. "Hello. Come in, come in," he said, smiling and getting up with his hand outstretched. When I introduced myself, he invited me to sit down and explain why I had come to see him. Feeling a little nervous, I got straight to the point, and it took only a few minutes to present the core of my proposal, explain that I was writing it as a case for change, and looking for support from the marine conservation community. Dr. Pauly listened, and after a few moments he said simply, "Yes, that's what should happen."

"Oh." I was surprised he agreed so easily. Wanting to be sure, I carried on, "Because protecting only patches of the ocean in marine reserves doesn't stop the root problem of bad practices, it just moves them to another place."

"You don't need to convince me," he said. "What you're proposing is what we should do." He paused and added, "It will be difficult to achieve—it could take decades—but it's what we should do. It makes sense."

Not only did he like the idea, he was happy to go on record as such, adding, "yes, it's radical—too radical for some—but that's how we need to be now."

Not long into our conversation Dr. Pauly asked whether I knew about a similar proposition concerning fisheries made by Dr. Carl Walters, one of his colleagues at the university. He had put it forward as a contribution to a book called *Reinventing Fisheries Management*, in a chapter titled "Designing Fisheries Management Systems That Do Not Depend upon Accurate Stock Assessment." Dr. Pauly took the book down from the shelf and passed it to me. I hadn't known about the proposition, and I was intrigued. The abstract of Dr. Walters's contribution begins,

> Fisheries management systems are coming to rely more on quota and allowable catch regulation, which in turn depends for success on accurate stock size assessments. But we cannot provide accurate assessments for most fish stocks, and probably never will. Instead of treating the seas as open to fishing with small exceptions (marine refugia), we will only safely limit harvest rates if we reverse this view and treat the seas as closed to fishing with small exceptions (limited fishing areas and times).[5]

In other words, he was proposing that seas should be closed to fishing by default, with some exceptions where they could be open for moderate and rational fishing—in much the same way that the whole ocean could be protected by default, with some exceptions— where it could be open for moderate and rational use. Apparently Dr. Walters has since backtracked on his "closed-unless-open-approach" to fisheries management, and it isn't clear why. Up against the vested interests of industry, perhaps he thought it would be too difficult to realize in practice.

The discussion meandered on from Dr. Walters to marine reserves and then to Charles Darwin (because I had noticed three shelves beside me, bowed under the weight of books either by or about Darwin). Dr. Pauly seemed content to chat for as long as I wanted about jellyfish, the breakup of Pangaea, toothless laws, high seas fishing, satellite tracking, how fish gills work, and Darwin. Over the next three weeks I returned several times to talk more with him and to meet members of his team and learn something about their work: Dr. Deng Palomares (senior scientist and project manager of the Sea Around Us); William Cheung (an associate professor working on climate change), and Dr. Rashid Sumaila (professor and director of the Fisheries Economics Research Unit). When it was time to leave Canada and return home to England I felt a little sad, especially at saying farewell to Ellie, but

it had been an unexpected treat to meet Daniel Pauly and his colleagues at UBC.

I was in a bookstore in the departure lounge with an hour to kill before
boarding my flight home from Vancouver. Flicking through a book on
diving, I came upon a stunning image of an octopus. The Caribbean
reef octopus pictured was a surprisingly bright blue, but I knew they
weren't always so vivid. They can alter their color to reflect certain
moods or, more often, as camouflage to hide from predators. It's also
useful for hunting, letting the animal merge into the background and
surprise its prey. This species is said to have a particularly wide self-
pigmentation palette ranging from dull grays and browns to greens,
blues, and even crimson. They can also change their skin texture from
smooth to lumpy. Octopuses' skins contain chromatophores, special-
ized light-reflecting cells that alter their skin color and even create
patterns to help them blend in. Add to that the ability to modify their
body shape, and they can be nearly impossible to see. You have to be
pretty smart to be aware of the shade and texture of what's behind or
beneath you and make yourself look almost the same, and octopuses
are among the most intelligent creatures living in the sea.

A lovely story from New Zealand's National Aquarium illustrates
just how clever and resourceful these animals are. A common New
Zealand octopus had been given to the aquarium by a fisherman who
had caught it in nearby Napier Bay. People who spend time with octopuses, in captivity or in the wild, note their individuality, each with
a particular personality. The little octopus in the aquarium acquired
the name Inky, and with such an inquisitive and engaging nature
quickly became one of the staff's favorite residents. Then, in the summer of 2016, Inky made a daring escape to freedom. One night when
all was quiet, he managed to move the lid of his enclosure slightly
aside and slip through the opening. He slid down the side of the tank,
crossed the floor, squeezed into a narrow drainpipe leading to the
sea, and made his way home to the ocean. The next morning, when
staff noticed he was missing, Inky's escape route was given away by

a wet trail running from the enclosure to the pipe. Rob Yarrell, the aquarium's manager, was quoted in *National Geographic News*: "We'll miss him, but we hope he does well in his new life."

Caribbean reef octopuses live on coral reefs and in sea-grass beds throughout shallow tropical seas in the western Atlantic. The one pictured in the diving book was spotted around the island of Ambergris Caye, about ten miles off the coast of Belize, in waters known to be among the world's most life-filled, pristine and idyllic in every way. I have heard and read about other such blissful places where the sea is crammed with life, such as Linapacan Island in the Philippines, São Tome and Principe, Baa Atoll in the Maldives, and Blue Corner in Palau. I'd love to visit one of these islands, and I imagine sailing there on a weather-beaten schooner, running before the wind, cutting a white furrow through the dark sea. In a different life perhaps. Another enticing destination is Juan Fernández, a group of three islands that rise dramatically through the surface of the Pacific, four hundred miles to the west of Chile. The largest of the islands, Robinson Crusoe, is named after Daniel Defoe's fictional character who was marooned on a desert island. Crusoe was supposedly based on the real Alexander Selkirk, after whom the second-largest island is named (Santa Clara is the third and smallest island of the archipelago). During an expedition across the Pacific, Scottish sailor and privateer Selkirk fell out with the ship's commander about the vessel's seaworthiness. He insisted on being left ashore on the uninhabited island that now bears his name. Little did he know he would be stranded there alone for the next four and a half years until a passing ship rescued him. On Selkirk's return home, his widely publicized story of survival inspired Defoe to create Selkirk's alter ego in his famous novel published in 1719. There is some doubt whether Alejandro Selkirk Island is the actual location where the self-styled castaway was left. If not, it's an engaging tale anyway.

The Juan Fernández archipelago is a watery wonderland in the South Pacific, home to an unusual mix of tropical, subtropical, and temperate species. Consequently there is an explosion of life under the waves. The Juan Fernández rock lobster and Juan Fernández fur seal are unique to this area, and green sea turtles, black sea turtles, bottlenose dolphins, humpback whales, sperm whales, blue whales, and other animals migrate through the islands' surrounding sea. Recognizing the irreplaceable value of its biodiversity, the Chilean government designated 187,000 square miles of ocean around the islands as a marine protected area in 2017. For the eight hundred people who live on these islands "lobster is life," and for centuries the remoteness of Juan Fernández not only kept rock lobster numbers high but protected all marine life. In the 1940s, though, as the modern world encroached, bringing the threat of overexploitation, the islanders agreed on self-imposed progressive conservation measures to protect this vital resource. The community decided to close the fishery for four months each year when the lobsters are breeding and spawning; all lobsters sold must be at least eleven inches long; any female with eggs has to be returned to the sea; and fishing rights cannot be bought or sold but are passed down through the families. Fishing pressure and fishing methods have stayed largely unchanged in Juan Fernández for generations, and more than half a century of considered and controlled fishing practice has paid dividends: the lobster fishery is thriving—which is another good news story to fend off the bad.

A few years have passed since the idea of protecting all of all the oceans blew in that night when I was camping on a Dorset hillside. In the short time since, we are already nearer that eventuality, and as each year passes it becomes less an idealistic notion and more an achievable strategy. The incremental shift from "ocean exploitation first" to "ocean protection first" has begun, and it is happening on a few fronts. I would say it started in the South Pacific island nation of Palau. The waters surrounding its 340 islands are regarded as among the most spectacular ocean environments on Earth. In 2015 the Palauan government made an inspirational announcement. With a unanimous vote in the House of Delegates and the approval of the Senate, it passed the Palau Marine Sanctuary Act, designating the nation's entire two-hundred-mile exclusive economic zone as a marine sanctuary. That means 193,000 square miles of the ocean is protected, home to 1,300 species of fish and 700 species of coral. Of waters under Palau's jurisdiction, 80 percent are fully protected as a reserve and the remaining 20 percent are accessible only to local fishers to carry out responsible, low-impact fishing. There will be no industrial fishing, drilling for oil, or mining within Palau's exclusive economic zone. "Today is a historic day for Palau, proving that a small island nation can have a big impact on the ocean," announced the country's president, Tommy E. Remengesau Jr. "Island communities have been among the hardest hit by the threats facing the ocean. Creating this sanctuary is a bold move that the people of Palau recognize as essential to our survival."[6]

With limited resources, enforcing the law over so large an area of the ocean is not easy. But when four Vietnamese ships were caught fishing illegally, Palau showed it meant business. The crews were arrested, the catch was confiscated, and the vessels were set on fire. President Remengesau declared, "We wanted to send a very strong message. We will not tolerate any more these pirates who come and steal our resources. Captains will be prosecuted and jailed. Boats will be burned. Nothing will be gained from poaching in Palau. From one fisherman to another, respect Palau."[7]

The concept of blanket protection by default, while permitting rational and respectful use, has become a reality around the archipelagic paradise of Palau. Islanders are already reaping the rewards. Less than two years after the sanctuary was created, scientists from the University of Hawaii found that the protected waters had twice as many fish as the unprotected waters and five times as many predatory fish.[8] In 2017 the Cook Islands followed Palau's lead when Parliament passed a bill to establish the Marae Moana Marine Park, a multiple-use marine protected area extending across the whole EEZ and covering 762,000 square miles, making it the world's largest protected area. "Mankind only has one earth, one atmosphere, and one global ocean. One last chance to save it all for future generations," said Prime Minister Henry Puna of the Cook Islands at the United Nations Ocean Conference in 2017.[9]

Meanwhile, shortly before Palau and the Cook Islands made the move to protect their entire ocean realms, another potential protective strategy—in a similar vein—was emerging elsewhere. It was first

presented by marine biologist Crow White and economist Christopher Costello in a paper titled "Close the High Seas to Fishing?," in
which they went on to answer their own question with a resound-
ing yes. With the high seas closed to commercial fishing, White and
Costello predicted that fish stocks would increase significantly, not
only out in the open oceans but within countries' EEZs too. As a result, yields would rise, and for some commercial species profits could
double.[10] About this time, fisheries scientists Rashid Sumaila and his
colleagues at the University of British Columbia were thinking the
same way, and the following year they published a paper expanding
on why closing the high seas to industrial fishing was a good idea. It
would create a safe haven for migratory species and a valuable "world
fish bank" in the global commons, where commercial stocks could
recover and would then spill over EEZ boundaries, boosting inshore
fish catches. That in turn would increase revenue for coastal states
and help even out the present inequalities caused by the bulk of high
seas fish catches' going to countries that can afford to subsidize their
fleets.

Currently a handful of countries are taking an unfair share of this
valuable commons resource — fish in the high seas — which isn't working out well for the rest of us, particularly those living in coastal states,
who rely on fish as their main source of protein. If fuel subsidies
ended tomorrow, most high seas fishing would not be economically
viable and would cease.[11] The expected benefits of the high seas' being
closed to industrial fishing would also take us a long way toward fulfilling the Law of the Sea's conservation obligations. The arguments
for closure and the data supporting them are compelling, and the advantages would be wide-ranging and long-lasting socially, economically, and environmentally. Although the principle of freedom of the
seas would be curtailed, freedom to fish high seas resources would
not, as EEZ waters would be restocked with high-value migratory
species like tuna. Not to include closing the high seas to industrial
fishing in the United Nations BBNJ treaty is to miss a great opportunity, not least because chances to amend the treaty come around so
infrequently. Would we really have to wait many blue moons until the
next time the UN decides to update the treaty? Apparently not. Article

313 of the Law of the Sea, titled "Amendment by Simplified Procedure," enables amendments to be agreed on without the need to con-
vene a conference of all the parties.

When I began the research for this book, conservation organizations were pressing governments to safeguard more of the global
ocean. The coverage they sought varied, but a common call was for
10 percent of it to be officially protected (which is also the Convention on Biodiversity's marine biodiversity target). The united clamor
of science and conservation is now for 30 percent of the global ocean
to be protected by 2030. In 2019, collaborating with the University
of Oxford and the University of York, Greenpeace launched the 30
percent proposal in the report "30X30: A Blueprint for Ocean Protection."[12] Significantly (and oddly, considering its poor record of protection in mainland waters) the British government joined the rallying
cry. "The Government of the UK has called for protecting thirty percent of the world's oceans by 2030. UK Environment Minister Thérèse
Coffey urged other countries to designate thirty percent of the world's
oceans as marine protected areas (MPAs) to address challenges posed
by climate change, plastic pollution and overfishing."[13] A robust,
large-scale system of control and enforcement will be needed to properly protect 30 percent of the world's oceans. But if you can create that
to safeguard a third of the ocean, why stop there? Why not safeguard
the whole lot?

I started this venture by asking if it is feasible to the turn the paradigm of the sea on its head and protect all oceans from the outset,
allowing only respectful use of marine resources. I found there are a
wealth of programs and projects up and running to counter ocean decline, with more on the way. They include a growing number of more
advanced fisheries regimes; cross-government cooperation to protect the Southern Ocean and its wildlife; citizen-driven campaigns to
clean polluted shores and seas; a wide range of dynamic NGO initiatives; an ever greater proportion of the sea protected in reserves and
protected areas; a thriving movement of grassroots fisheries management leapfrogging from one continent to another; and further protective national and international legislation in the making. I also came
across some inspiring individuals running creative campaigns while

bemoaning the lack of political courage to do more and lamenting governments' continuing deference to the very industries and institutions that are damaging and destroying the natural world.

Up to now the Law of the Sea, the public trust doctrine, the Convention on Biological Diversity, and other international agreements have been failing to safeguard the sea, for similar reasons. Authorities frequently overlook inconvenient laws intended to protect nature and shirk their duty to manage the use of natural resources well. This is double-whammy bungling. By breaching both treaty law and customary law, governments and leaders are allowing the ruin of marine environments while also denying citizens' rights to all that oceans provide, not least food and employment for millions of people.

Few people have heard of the public trust doctrine, but that is changing. Certainly, where it's embedded in the constitution the doctrine has the potential to make the state do more for nature. Whether they use customary law or treaty law, groups of people are increasingly using one law to make their governments comply with another (primarily regarding cutting CO_2 emissions). Any success in combating climate change will be beneficial to the health of the oceans too, by keeping water temperatures stable, slowing ice-cap melt, and moderating ocean acidification. The question is whether the legal route can rescue the ocean specifically, in the same way it is beginning to do in campaigns to reduce CO_2 emissions (provided that governments do make the changes demanded by a court ruling). As happened with the Supreme Court ruling in the Urgenda case, if courts of law can make governments act to reduce greenhouse gases and force domestic regulations to come into line with international agreements, they should be able to do the same for a country's legal obligation to protect the oceans. Could courts rule against the state's spending public money to perpetuate overfishing, force authorities to introduce sustainable fisheries policies, and enforce laws meant to prevent pollutants from entering the sea?

A number of international laws are already in place to protect the marine environment. But what's the point of states agreeing to laws, signing up, and then taking little or no notice of them? Such laws, if followed, would address a range of serious environmental threats facing coasts and oceans across the world. Conservation NGOs, opposition parties, and civil society groups could better exploit so much negligence and contravention and take offending authorities to court for acting outside the law. Now I'm envisaging a citizens vs. government legal challenge to safeguard the sea.

Until we find a better way to govern societies, and for as long as we are subject to hierarchical political systems, if I were granted one wish for the sea it would be to have a very different style of leadership across the world, especially in positions of the greatest power. We would have political leaders who work together in a meaningful and intelligent way to find solutions to today's environmental problems and those anticipated in the future—and although some do work this way, many do not. They would be well-informed men and women who engender trust in the public, with a broad and balanced understanding of the issues they are dealing with—abandoning ego, catchphrase thinking, and divisive rhetoric in favor of tolerance and truth. They would be willing to listen to sound advice and be prepared to compromise to reach agreements for the common good. Most of all, they would be leaders with compassion and integrity. Dream on, you say. How do we get leaders like that? Some would argue that we shouldn't have leaders at all. Maybe so, but in the meantime, we clearly need better ones. In 2019, at the UN Climate Action Summit in New York, young Swedish activist Greta Thunberg made a powerful and impassioned address to world leaders, berating them for forsaking her generation by not doing more to prevent climate change and an ever greater loss of biodiversity. "We are at the beginning of a mass extinction," she told them, "and all you can talk about is money and fairy tales of eternal economic growth. How dare you!?"

Creating marine protected areas and reserves is widely regarded as the silver bullet of ocean conservation, when in truth it is an exercise in damage limitation that enables cruel and harmful commercial practices to continue over large parts of the world's oceans. United, multinational political commitment to modernize and implement existing international law can generate a wholly different, more enlightened understanding of the ocean and its wildlife. All serious sea-based problems, together with those originating on land, can be tackled by taking a precautionary approach to everything people do with the sea. It's a strategy of "first do no harm" that insists on using all marine resources responsibly. We need to be smarter, and let's hurry up about it. A resilient natural world is central to life itself and should be central to policy making at the highest level, bringing our attitude to nature out of the Dark Ages and into the twenty-first century. Our grandchildren and great-grandchildren may regard recent generations as weak and inhumane, ruthless, or willfully ignorant, with upside-down reasoning that values money more than life. And perhaps they will look further back and think the way we do now about a society that enabled millions of free men, women, and children to be captured, transported across oceans in brutal conditions, and traded as mere commodities with the same tacit acceptance of the unacceptable.

Essentially, the long-term way to tackle the environmental troubles at sea is through education, promoting better understanding and sharing knowledge—at every level of society, in every sector, in every land. When people understand, they'll care, and when they care about something they'll look after it. In the era of common sense, equitable and rational use of the sea's resources will help to preempt conflict between states and encourage international peace. Ocean waters will become cleaner, there will be plenty of fish to eat and to sell, damaged undersea habitats will recover, wildlife will come back, wholesale cruelty will end, and one day people will wonder why the present situation was ever allowed to continue so long.

— fishing

Acknowledgments

Many people helped me in writing this book, and I welcome the opportunity to thank them: James Attlee for setting the project in motion and for his support throughout, and Christie Henry, former science editor at the University of Chicago Press, for inviting me to turn an idea into a book. For their interest and input I thank all my family, my friends—especially Clare Hartland, Penny Rogers, and Martin Tillett—and my daughters, Ellie Sheldrake and Lydia Sheldrake. I'm also grateful to members of the marine conservation organization Marinet, particularly Stephen Eades and David Levy, and the team at the University of Chicago Press for guiding me along the unfamiliar trail of writing a book: Alice Bennett, Scott Gast, Nick Lilly, Michaela Luckey, Christine Schwab, Alan Thomas, Rachel Kelly Unger, and the rest.

I have met, talked with, emailed, and been helped by many specialists and campaigners and others who contributed in some way, and I thank them all for their expertise, time, and goodwill: Stephen O. Andersen, Mark Belchier, Chris Bone, Jon Paul Brooker, Suely Carvalho, William Cheung, Neville Copperthwaite, Emily Corcoran, Steve Downey, Natalie Fee, Jim Green, John Guilfoyle, Polly Higgins, Ray Hilborn, Gary Jarvis, David Johnson, Jeroen Jongejans, Tommy Koh, Jason Link (NOAA), Jojo Mehta, Chris Millman, George Monbiot, Charli Moore, Geir Ottersen, Deng Palomares, Daniel Pauly, Sharon Redmayne, Steve Rocliffe, Charles Secrett, Emily Shirley, James Simpson (MSC), Don Staniford, Rashid Sumaila, Mike Swindells, Stephen Taylor, Oliver Tickell, Sofia Tsenikli, Catherine Wallace, and Chris Warren (and forgive me if I haven't named everyone).

Notes

Chapter One

1. G. E. Fenton, S. A. Short, and D. A. Ritz, "Age Determination of Orange Roughy, *Hoplostethus atlanticus* (Pisces: Trachichthyidae) using ^{210}Pb$:^{226}$Ra Disequilibria," *Marine Biology* 109 (1991): 197–202.
2. UN Food and Agriculture Organization, Fisheries and Aquaculture Department, "Species Fact Sheets: *Hoplostethus atlanticus* (Collett, 1889)," 2019, http://www.fao.org/fishery/species/2249/en.
3. UN Division for Ocean Affairs and the Law of the Sea, "World Oceans Day—the Importance of a Healthy Ocean," photo exhibition for World Oceans Day 2013, https://www.un.org/Depts/los/reference_files/WODreferenceMaterials/WOD 2013_PhotoExhibition.pdf.

Chapter Two

1. UN Division for Ocean Affairs and the Law of the Sea, "United Nations Convention on the Law of the Sea," December 10, 1982, https://www.un.org/depts/los/convention_agreements/texts/unclos/unclos_e.pdf.
2. Alejandra Borunda, "Ocean Acidification, Explained," *National Geographic* online, August 7, 2019, https://www.nationalgeographic.com/environment/oceans/critical-issues-ocean-acidification/.
3. United Nations Treaty Collection, Certified True Copies of Multilateral Treaties Deposited with the Secretary-General, chapter 21, item 7, "Agreement for the Implementation of the Provisions of the United Nations Convention on the Law of the Sea of 10 December 1982 relating to the Conservation and Management of Straddling Fish Stocks and Highly Migratory Fish Stocks," August 4, 1995, https://treaties.un.org/doc/Treaties/1995/08/19950804%2008-25%20AM/Ch_XXI_07p.pdf.
4. Convention on Biological Diversity, "Article 8. In-situ Conservation," March 30, 2007, https://www.cbd.int/convention/articles/default.shtml?a=cbd-08.
5. United Nations, "United Nations Framework Convention on Climate Change,"

1992, https://unfccc.int/files/essential_background/background_publications
_htmlpdf/application/pdf/conveng.pdf.

6. Garrett Hardin, "The Tragedy of the Commons," *Science* 162, no. 3859 (1968): 1243–48, https://doi.org/10.1126/science.162.3859.1243.

Chapter Three

1. Laura Hampton, "Never-before-Seen Sea Creatures Filmed in World's Deepest Abyss," *New Scientist*, July 12, 2016, https://www.newscientist.com/article/2096 973-never-before-seen-sea-creatures-filmed-in-worlds-deepest-abyss/.

2. Daniel Pauly, "Anecdotes and the Shifting Baseline Syndrome of Fisheries," *Trends in Ecology and Evolution* 10, no. 10 (1995): 430.

3. Steve Mackinson, "Representing Trophic Interactions in the North Sea in the 1880s Using the Ecopath Mass-Balance Approach," in *Fisheries Impacts on North Atlantic Ecosystems: Models and Analyses*, ed. Sylvie Guénette, Villy Christensen, and Daniel Pauly, 35–98, Fisheries Centre Research Reports, vol. 9, no. 4 (Vancouver: Institute for the Oceans and Fisheries, University of British Columbia, 2001).

4. Fiona Harvey, "North Sea Cod at Critically Low Levels, Study Warns," *Guardian*, June 29, 2019, https://www.theguardian.com/environment/2019/jul/29/north -sea-cod-at-critically-low-levels-study-warns.

5. Greenpeace, "The Collapse of the Canadian Newfoundland Cod Fishery," May 8, 2009, https://www.greenpeace.org/archive-seasia/ph/What-we-do/oceans/sea food/understanding-the-problem/overfishing-history/cod-fishery-canadian/.

6. Sea Around Us, "Pauly and Zeller Explain the Making of the 'Sea Around Us' Database," September 20, 2018, http://www.seaaroundus.org/pauly-and-zeller -explain-the-making-of-the-sea-around-us-database/.

7. Food and Agriculture Organization of the United Nations, "Illegal, Unreported and Unregulated (IUU) Fishing," http://www.fao.org/iuu-fishing/en/, accessed February 6, 2020.

8. E. Pikitch et al., *Little Fish, Big Impact: Managing a Crucial Link in Ocean Food Webs* (Washington, DC: Lenfest Ocean Program, 2012), https://www.lenfest ocean.org/en/news-and-publications/published-paper/little-fish-big-impact-a -report-from-the-lenfest-forage-fish-task-force.

9. Cyanide fishing is widely practiced in Southeast Asia even though it is illegal in most countries of the region. Divers squirt a solution of cyanide tablets and seawater around a coral reef to stun the fish so they can easily catch them. It is practiced primarily to capture live reef fish for the aquarium markets of Europe and North America and for high-end specialty restaurants in Asian cities.

10. Dynamite fishing is self-explanatory. Jani Actman sums it up in two short but disturbing sentences: "Some dynamite and a plastic bottle. That's all it

takes for a fisherman to kill hundreds of fish and transform thriving coral reefs into rubble in a matter of seconds." Jani Actman, "Watch Fishermen Bomb Their Catch Out of the Water," *National Geographic*, June 3, 2016, https://www .nationalgeographic.com/news/2016/06/blast-fishing-dynamite-fishing -tanzania/.

11. Common in Southeast Asia, especially in the Philippines, muro ami fishing involves divers' pounding and smashing coral reefs with heavy stones or cement blocks. This scares fish out of the shelter of the coral, and they swim into a large net. Besides destroying precious coral reefs, muro ami unfairly exploits children, because the divers are often young boys.

12. In shark finning fishers catch sharks, cut off their fins, and throw the animals overboard to die—all to supply the demand for shark-fin soup. Estimates of the number of sharks killed every year are from seventy to one hundred million. It is a shockingly brutal and cruel practice. Finning is also insanely wasteful, and it seriously disrupts the ecological balance of marine life because sharks are essential for keeping the populations of other species in check.

13. Longlines have a main line with individual lines branching off into the water column or on the seafloor. The branch lines are spaced out with baited hooks attached. Longlines in high seas fisheries can extend sixty miles and have thousands of hooks. Longlining causes significant numbers of bycatch deaths, particularly of pelagic sharks (inhabiting midwater to surface zones), turtles, and seabirds.

14. Elizabeth Wilson, "Fishing Subsidies Are Speeding the Decline of Ocean Health," Pew, July 19, 2018, https://www.pewtrusts.org/en/research-and-analy sis/articles/2018/07/19/fishing-subsidies-are-speeding-the-decline-of-ocean -health.

15. See European Commission, European Maritime and Fisheries Fund, https://ec .europa.eu/fisheries/cfp/emff_en, accessed February 6, 2020.

16. Christopher Pala, "Billions in Subsidies Prop Up Unsustainable Overfishing," IPS News Agency, November 8, 2012, http://www.ipsnews.net/2012/11/billions -in-subsidies-prop-up-unsustainable-overfishing/.

17. Jeff Barger, "MSA: 40 Years of Rebuilding Fishing Communities," *Ocean Currents* (blog), Ocean Conservancy, April 13, 2016, https://oceanconservancy.org /blog/2016/04/13/fish-town-usa/.

18. Barger, "MSA."

19. Environmental Investigation Agency and Animal Welfare Institute, "Commercial Whaling: Unsustainable, Inhumane, Unnecessary," September 2018, https:// awionline.org/sites/default/files/press_release/files/AWI-IWC-report-final -2018.pdf.

20. Jennifer Lonsdale, quoted in Rachael Bale, "Norway's Whaling Program Just Got Even More Controversial," *National Geographic* online, March 31, 2016,

https://www.nationalgeographic.com/news/2016/03/160331-norway-minke-whaling-fur-farms/.

21. J. Roman et al., "Whales as Ecosystem Engineers," *Frontiers in Ecology and the Environment* 12, no. 7 (2014): 377–85, http://cetus.ucsd.edu/sio133/PDF/Roman%20et%20al%202014%20Frontiers%20in%20Ecology%20&%20the%20Environment.pdf.

22. Sandra Altherr, Kate O'Connell, Sue Fisher, and Sigrid Luber, "Frozen in Time: How Modern Norway Clings to Its Whaling Past," a report by Animal Welfare Institute, Ocean Care, and Pro Wildlife, 2016, https://awionline.org/sites/default/files/uploads/documents/AWI-ML-NorwayReport-2016.pdf.

23. Tony Pitcher and Daniel Pauly, "Rebuilding Ecosystems, Not Sustainability, as the Proper Goal of Fishery Management," in *Reinventing Fisheries Management*, ed. Tony J. Pitcher, Paul J. B. Hart, and Daniel Pauly, 311–29, Fish and Fisheries Series 23 (Dordrecht: Kluwer, 1998).

24. Derek Staples and Simon Funge-Smith, "Ecosystem Approach to Fisheries and Aquaculture: Implementing the FAO Code of Conduct for Responsible Fisheries," Food and Agriculture Organization, 2009, http://www.fao.org/3/i0964e/i0964e00.htm.

25. The 1995 Code of Conduct for Responsible Fisheries is a set of principles and guidelines on best practice in the fishing industry that was agreed to by over 170 countries at a meeting of the United Nations Food and Agriculture Organization in 1995. Its purpose is to encourage responsible and sustainable management and to conserve biodiversity and abundance.

26. Stephen Nicol and Yoshinari Endo, "Krill Fisheries of the World," FAO Fisheries and Aquaculture Department, 1997, http://www.fao.org/3/W5911E/w5911e07.htm#b1-4.1%20Antarctic%20krill%20(Euphausia%20superba).

27. A. Atkinson et al., "A Re-appraisal of the Total Biomass and Annual Production of Antarctic Krill," *Deep-Sea Research* 56 (2009): 727–40.

Chapter Four

1. E. Lewis, ed., *Chronicles of the Sea* (1898).

2. Hugo Grotius, *Mare Liberum* (1609).

3. World Wildlife Fund, "Bycatch," https://www.worldwildlife.org/threats/bycatch, accessed February 6, 2020.

4. Orea Anderson et al., "Global Seabird Bycatch in Longline Fisheries," *Endangered Species Research* 14 (2011): 91–106.

5. Martin Fowlie, "Don't Let This Be the Last Post for Albatross," press release, Royal Society for the Protection of Birds, December 8, 2017, https://www.rspb.org.uk/about-the-rspb/about-us/media-centre/press-releases/dont-let-this-be-the-last-post-for-albatross/.

6. The common heritage of humankind (or mankind) states that the global com-

mons—areas beyond national territories such as the high seas—are humanity's common heritage and should not be overexploited by individual nations or corporations but are to be held in trust for the benefit of all and for future generations. The concept is attributed to the former Maltese ambassador to the United Nations, Arvid Pardo.

7. "Net Positive," *Economist*, July 14, 2016.
8. Andrew Jillions, *Commanding the Commons: Constitutional Enforcement and the Law of the Sea* (Cambridge: Cambridge University Press, 2015).
9. Deep Sea Conservation Coalition, "New Zealand Trawl Fleet to Continue Destruction of Deep-Sea Ecosystems in South Pacific on the High Seas," January 28, 2019, http://www.savethehighseas.org/2019/01/28/new-zealand-trawl-fleet-to-continue-destruction-of-deep-sea-ecosystems-%E2%80%A8in-south-pacific-on-the-high-seas/.
10. Amanda Nickson, "New Science Puts the Decline of Pacific Bluefin at 97.4 Percent," Pew, April 25, 2016, https://www.pewtrusts.org/en/research-and-analysis/articles/2016/04/25/new-science-puts-decline-of-pacific-bluefin-at-974-percent.
11. World Wildlife Fund, "Unacceptable Rise in Catch Quota for Bluefin Tuna! WWF Protests," November 21, 2017, http://mediterranean.panda.org/news/?uNewsID=317090.
12. Tony Long, "Global Fishing Watch in 2018—the Year in Transparency," Global Fishing Watch, December 18, 2018, https://globalfishingwatch.org/data/global-fishing-watch-2018-the-year-in-transparency/.
13. BBC News, "Ban on 'Brutal' Fishing Blocked," November 24, 2006, http://news.bbc.co.uk/1/hi/sci/tech/6181396.stm.
14. Callum Roberts, *The Ocean of Life: The Fate of Man and the Sea* (London: Allen Lane, 2012).
15. Dan Laffoley et al., "Eight Urgent, Fundamental and Simultaneous Steps Needed to Restore Ocean Health, and the Consequences for Humanity and the Planet of Inaction or Delay," *Aquatic Conservation: Marine and Freshwater Ecosystems*, July 23, 2019, https://onlinelibrary.wiley.com/doi/10.1002/aqc.3182.
16. Convention on Biological Diversity, "Target 6," October 2, 2012, https://www.cbd.int/sp/targets/rationale/target-6/.
17. United Nations Development Programme, "Goal 14: Life Below Water," https://www.undp.org/content/undp/en/home/sustainable-development-goals/goal-14-life-below-water.html, accessed March 18, 2020.

Chapter Five

1. United Nations Convention on the Law of the Sea, part XII, article 192.
2. Paul Krugman interviewed by Helen Lewis on *The Spark*, BBC Radio 4, January 1, 2020, https://www.bbc.co.uk/programmes/m000cp2w.

3. Martin Gilens and Benjamin Page, "Testing Theories of American Politics: Elites, Interest Groups, and Average Citizens," *Perspectives on Politics* 12, no. 3 (2014): 564–81.

4. Represent Us, "The Problem," https://act.represent.us/sign/the-problem, accessed February 6, 2020.

5. Luke Darby, "Billionaires Are the Leading Cause of Climate Change," *GQ Magazine*, December 22, 2018, https://www.gq.com/story/billionaires-climate -change.

6. Elinor Ostrom, *Governing the Commons: The Evolution of Institutions for Collective Action* (Cambridge: Cambridge University Press, 1990).

7. Global Ocean Commission, "The Future of Our Ocean: Next Steps and Priorities," report, February 2016, http://www.some.ox.ac.uk/wp-content/uploads/20 16/03/GOC_2016_Report_FINAL_7_3.low_1.pdf.

8. Elisabeth Mann Borgese, *The Oceanic Circle: Governing the Seas as a Global Resource* (Tokyo: United Nations University Press, 1998).

9. Charles Clover, *The End of the Line: How Overfishing Is Changing the World and What We Eat* (London: Ebury Press, 2004).

10. Callum Roberts, *The Unnatural History of the Sea* (Washington, DC: Island Press, 2007).

11. Crispin Dowler, "Revealed: The Millionaires Hoarding UK Fishing Rights," *Unearthed*, October 10, 2018, https://unearthed.greenpeace.org/2018/10/11/fish ing-quota-uk-defra-michael-gove/.

12. Pew, "Ending Overfishing in Northwestern Europe," https://www.pewtrusts.org /en/projects/ending-overfishing-in-northwestern-europe, accessed February 6, 2020.

13. Commission for the Conservation of Antarctic Marine Living Resources, "CCAMLR Strengthens Marine Conservation in Antarctica," November 2, 2012, https://www.ccamlr.org/en/organisation/ccamlr-strengthens-marine-conserva tion-antarctica-1.

14. Commission for the Conservation of Antarctic Marine Living Resources, "Updated Calculator Provides Times for Nautical Twilight, Sunrise and Sunset," September 8, 2014, https://www.ccamlr.org/en/news/2014/updated-calculator -provides-times-nautical-twilight-sunrise-and-sunset.

15. Callum Roberts, *The Ocean of Life: The Fate of Man and the Sea* (London: Allen Lane, 2012).

16. Interpol, "Fisheries Crime," https://www.interpol.int/en/Crimes/Environmental -crime/Fisheries-crime, accessed February 6, 2020.

17. World Wildlife Fund, "What Americans and Europeans Spend on Ice Cream: $31 Billion; Global Cost of Creating Marine Parks to Protect the Oceans: $12–14 Billion," June 14, 2004, https://www.worldwildlife.org/press-releases/what-ame ricans-and-europeans-spend-on-ice-cream-31-billion-global-cost-of-creating -marine-parks-to-protect-the-oceans-12-14-billion.

18. Stockholm International Peace Research Institute, "World Military Spending Grows to $1.8 Trillion in 2018," April 29, 2019, https://www.sipri.org/media/pre ss-release/2019/world-military-expenditure-grows-18-trillion-2018.

Chapter Six

1. Roger Cox, interviewed in 2015 for Responding to Climate Change (RTCC), https://thinkprogress.org/in-landmark-case-dutch-citizens-sue-their-govern ment-over-failure-to-act-on-climate-change-e01ebb9c3af7/.
2. Martin Enserink, "In Surprise, Dutch Court Orders Government to Do More to Fight Climate Change," *Science*, June 24, 2015, https://www.sciencemag.org /news/2015/06/surprise-dutch-court-orders-government-do-more-fight-cli mate-change.
3. Backlight TV documentary by VPRO, http://www.revolutionjustified.org/roger -cox-author-of-revolution-justified/9-videos/28-vpro-tegenlicht.
4. Joseph Sax, "The Public Trust Doctrine in Natural Resource Law: Effective Judicial Intervention," *Berkeley Law Scholarship Repository* 68 (1969): 471–566.
5. Mary Christina Wood, *Nature's Trust: Environmental Law for a New Ecological Age* (Cambridge: Cambridge University Press, 2013).
6. Channel 4, "Saving Planet Earth: Fixing a Hole," 2018; and PBS, "Ozone Hole: How We Saved the Planet," 2018, https://www.pbs.org/show/ozone-hole-how -we-saved-planet/.
7. C. A. Downs et al., "Toxicopathological Effects of the Sunscreen UV Filter, Oxybenzone (Benzophenone-3), on Coral Planulae and Cultured Primary Cells and Its Environmental Contamination in Hawaii and the U.S. Virgin Islands," *Archives of Environmental Contamination and Toxicology* 70, no. 2 (2015): 265–88, https://doi.org/10.1007/s00244-015-0227-7.
8. Downs et al., "Toxicopathological Effects."
9. Ecocide Law, https://ecocidelaw.com/the-law/what-is-ecocide/, accessed February 6, 2020.
10. Marlene Moses, "Small Island States with Big Ocean Visions," Ocean Unite, https://www.oceanunite.org/small-island-states-big-ocean-visions/, accessed February 6, 2020.
11. Stop Ecocide, https://www.stopecocide.earth/, accessed February 6, 2020.

Chapter Seven

1. Ian Johnston, "Women Must Have Equality with Men to Save the Planet, Experts Say," *Independent*, April 21, 2017. https://www.independent.co.uk/environment /women-men-equality-save-planet-say-experts-a7693781.html.
2. Antonio Oposa Jr., TEDxManila, May 12, 2009, https://www.youtube.com/watch ?v=PPLLZX_bMdI.

3. Gerardo Ceballos, Paul R. Ehrlich, and Rodolfo Dirzo, "Biological Annihilation via the Ongoing Sixth Mass Extinction Signaled by Vertebrate Population Losses and Declines," *Proceedings of the National Academy of Sciences* 114 (30): E6089–E6096 (July 25, 2017), https://doi.org/10.1073/pnas.1704949114.

4. Damien Carrington, "Ocean Temperatures Hit Record High as Rate of Heating Accelerates," *Guardian*, January 13, 2020, https://www.theguardian.com/enviro nment/2020/jan/13/ocean-temperatures-hit-record-high-as-rate-of-heating -accelerates.

5. When rain washes agricultural fertilizers off the land and into rivers, lakes, and the sea, they enrich the waters with excess nutrients, mainly nitrogen and phosphorus. This encourages the growth of algae, which covers the water's surface and stops sunlight from penetrating, thereby preventing other aquatic plants from photosynthesizing, which has the effect of deoxygenating the water. Known as eutrophication, the process kills off underwater life beneath the algal canopy by a lack of oxygen in the water. Dead organisms sink to the bottom and break down, compounding the problem as decomposition uses up what's left of the oxygen. Eventually a "dead zone" is created. In August 2017 the American government agency the National Oceanic and Atmospheric Administration (NOAA) reported that a dead zone in the Gulf of Mexico was 8,776 square miles, *"the largest measured since dead zone mapping began there in 1985."* Globally, there are over four hundred dead zones, with a combined area covering approximately 92,600 square miles. R. J. Diaz and R. Rosenberg, "Spreading Dead Zones and Consequences for Marine Ecosystems," *Science* 321 (2008): 926–29.

6. Laura Parker, "To Save the Oceans, Should You Give Up Glitter?," *National Geographic*, November 2017.

7. UN Environment, "Our Planet Is Drowning in Plastic Pollution," https://www .unenvironment.org/interactive/beat-plastic-pollution/, accessed February 6, 2020.

8. J. Jambeck et al., "Plastic Waste Inputs from Land into the Ocean," *Science* 347, no. 6223 (2015): 768–71, https://doi.org/10.1126/1260352.

9. Commonwealth Clean Ocean Alliance, https://bluecharter.thecommonwealth .org/action-groups/marine-plastic-pollution/.

10. Robert Ferris, "Deep Sea Mining Company Reveals New Gear," CNBC, November 2015, https://www.cnbc.com/2015/11/11/worlds-first-deep-sea-mining-compa ny-reveals-new-gear.html.

11. Dan Laffoley et al., "Eight Urgent, Fundamental and Simultaneous Steps Needed to Restore Ocean Health, and the Consequences for Humanity and the Planet of Inaction or Delay," *Aquatic Conservation: Marine and Freshwater Ecosystems*, 2019, https://onlinelibrary.wiley.com/doi/10.1002/aqc.3182.

12. Greenpeace, "Four Reasons Why the International Seabed Authority Shouldn't Be Trusted to Protect Our Oceans," July 14, 2018, https://www.greenpeace.org

.uk/news/four-reasons-why-the-international-seabed-authority-shouldnt-be
-trusted-to-protect-our-oceans/.

13. World Wildlife Fund, "Farmed Salmon," https://www.worldwildlife.org/indus
    tries/farmed-salmon, accessed February 6, 2020.

14. Don Staniford of the Global Alliance Against Industrial Aquaculture, inter-
    viewed in the *Guardian*, 2017, https://www.theguardian.com/environment/20
    17/apr/01/is-farming-salmon-bad-for-the-environment.

15. Feedback, "Fishy Business: The Scottish Salmon Industry's Hidden Appetite
    for Wild Fish and Land," 2019, https://feedbackglobal.org/wp-content/uploads
    /2019/06/Fishy-business-the-Scottish-salmon-industrys-hidden-appetite-for
    -wild-fish-and-land.pdf.

16. Paul Greenberg, *Four Fish: A Journey from the Ocean to Your Plate* (Westminster,
    UK: Allen Lane, 2010).

17. Dr. Joseph Mercola, "Farmed Salmon = Most Toxic Food in the World," Organic
    Consumers Association, July 15, 2018, https://www.organicconsumers.org/news
    /farmed-salmon-toxic-flame-retardants.

18. Ronald Hites et al., "Global Assessment of Organic Contaminants in Farmed
    Salmon," *Science* 303, no. 5655 (2004): 226–29, https://doi.org/10.1126/1091447.

19. Don Staniford, pers. comm. See also Scottish Salmon Watch, https://scottish
    salmonwatch.org/about-us, accessed February 6, 2020.

20. Sea Shepherd Conservation Society is a direct action, nonprofit marine conser-
    vation organization that aims to stop illegal activities on the sea, notably whal-
    ing, seal hunting, and fishing. Founded by Paul Watson in 1977, it is best known
    for challenging Japanese whaling ships and trying to prevent them from oper-
    ating. Sea Shepherd will intercept the whalers with one of their ships and even
    ram them, though the organization is committed to a nonviolent approach.
    Other targets of their campaigns are the horrific annual mass slaughters of
    pilot whales in the Faroe Islands and of dolphins in Taiji, Japan. Sea Shepherd
    ships patrol and police protected areas as well, such as around the Galápagos
    Islands and in the Gulf of California. More recently they have been active in the
    waters off West Africa, to assist the authorities of countries including Senegal
    and Liberia with law enforcement at sea.

    West Africa is the region of the world most subject to illegal, unreported and
    unregulated (IUU) fishing, with an illegal catch estimated to cost West Afri-
    can countries between 608 million euros and 1.2 billion euros per year. IUU
    fishing in West Africa accounts for 37% of illegal (pirate) fishing globally.
    Senegal, with more than seven hundred kilometers of coastline and waters
    teeming with a diversity of species ranging from sharks, tuna and swordfish,
    is one of the African countries most targeted by foreign pirate fishing fleets
    from Europe, Asia and Russia. (Sea Shepherd UK, "Sea Shepherd Launches

'Operation Sunu Gaal,'" *Sea Shepherd News*, January 11, 2014, https://www
.seashepherd.org.uk/news-and-commentary/news/sea-shepherd-launches
-operation-sunu-gaal.html.)

Sea Shepherd has always claimed to be enforcing international laws to protect marine life, in the absence of adequate enforcement by national authorities.

Since 1977, Sea Shepherd has used innovative direct action tactics to defend,
conserve and protect the delicately-balanced biodiversity of our seas and enforce international conservation laws. From its earliest years, Sea Shepherd
has embraced the mandate of the United Nations World Charter for Nature
to uphold international conservation laws when nations can't . . . or won't.
(Sea Shepherd, "Our Mission," https://www.seashepherdglobal.org/who-we
-are/our-mission/, accessed February 6, 2020. Sections 21–24 of the charter
provide authority to individuals to act on behalf of and enforce international
conservation laws.)

Chapter Eight

1. Charli Moore, "The Poor Knights Islands Marine Reserve," Wanderlusters, 2013,
    https://wanderlusters.com/poor-knights-islands-marine-reserve/.
2. J. S. Ault et al., "Building Sustainable Fisheries in Florida's Coral Reef Ecosys-
    tem: Positive Signs in the Dry Tortugas," *Bulletin of Marine Science* 78, no. 3
    (2006): 633–54.
3. S. A. Murawski et al., "Large-Scale Closed Areas as a Fishery-Management
    Tool in Temperate Marine Systems: The Georges Bank Experience," *Bulletin of
    Marine Science* 66, no. 3 (2000): 775–98.
4. J. García-Charton et al., "Effectiveness of European Atlanto-Mediterranean
    MPAs: Do They Accomplish the Expected Effects on Populations, Communities
    and Ecosystems?," *Journal for Nature Conservation* 16, no. 4 (2008): 193–221.
5. S. E. Lester et al., "Biological Effects within No-Take Marine Reserves: A Global
    Synthesis," *Marine Ecology Progress Series* 384 (2009): 33–46, https://doi.org/10
    .3354/meps08029.
6. Lyme Bay Fisheries and Conservation Reserve, https://www.lymebayreserve.co
    .uk/science/.
7. The White House, Office of the Press Secretary, "Presidential Proclamation—
    Pacific Remote Islands Marine National Monument Expansion," September 25,
    2014, https://obamawhitehouse.archives.gov/the-press-office/2014/09/25/presi
    dential-proclamation-pacific-remote-islands-marine-national-monumen.
8. UK Government, "Introducing the Blue Belt Programme," 2017, https://assets
    .publishing.service.gov.uk/government/uploads/system/uploads/attachment
    _data/file/662392/27_OCT_Introducing_Blue_Belt_FINAL-_updated1.pdf.

9. Chelsea Harvey, "In Historic Agreement, Nations Create the World's Largest Marine Reserve in Antarctica," *Washington Post*, October 27, 2016, https://www.washingtonpost.com/news/energy-environment/wp/2016/10/27/in-historic-agreement-nations-forge-the-worlds-largest-marine-reserve-in-antarctica/?noredirect=on.

10. The Benguela Current Large Marine Ecosystem (BCLME) is an area of extremely productive waters rich in marine species. The Benguela ocean current flows northward along the coast of West Africa, from the continent's southern tip almost up to the equator, then veers west. Recognizing the biodiverse value of these waters, the governments of Angola, Namibia, and South Africa set up the Benguela Current Commission and agreed on a joint program to manage the region's resources in a responsible way.

11. Whitley Fund for Nature, "Alifereti Tawake—Everlasting Marine Resources for our Descendants, Fiji," https://whitleyaward.org/winners/everlasting-marine-resources-fiji/, accessed February 6, 2020.

12. Blue Ventures, "Western Indian Ocean Communities Playing Vital Role in Conservation," July 23, 2014, https://blueventures.org/west-indian-ocean-communities-playing-vital-role-in-conservation/.

13. R. Vave et al., "The Effectiveness of Locally Managed Marine Areas in Fiji," University of the South Pacific, Conference Proceedings, June 2, 2014, http://repository.usp.ac.fj/7416/.

Chapter Nine

1. Dirk Notz and Julienne Stroeve, "Observed Arctic Sea-Ice Loss Directly Follows Anthropogenic CO_2 Emission," *Science* 354, no. 6313 (2016): 747–50, https://doi.org/10.1126/science.aag2345.

2. The World Bank, "CO_2 Emissions (Metric Tons per Capita)," 2019, https://data.worldbank.org/indicator/EN.ATM.CO2E.PC.

3. World Energy Council, "World Energy Resources: Marine Energy," 2016, https://www.marineenergywales.co.uk/wp-content/uploads/2016/01/World-Energy-Council-Marine-Energy-Resources-2016.pdf.

4. OTEC News, "What Is OTEC," http://www.otecnews.org/what-is-otec/, accessed February 6, 2020.

5. Patrick Greenfield, "Top Investment Banks Provide Billions to Expand Fossil Fuel Industry," *Guardian*, October 13, 2019. https://www.theguardian.com/environment/2019/oct/13/top-investment-banks-lending-billions-extract-fossil-fuels.

6. ParliamentLive.tv, Treasury Committee, October 15, 2019, https://www.parliamentlive.tv/Event/Index/a22c3906-54d0-43fe-bbc1-03f0284e5491.

7. J. Poore and T. Nemeck, "Reducing Food's Environmental Impacts through Pro-

ducers and Consumers," *Science* 360, no. 6392 (2018): 987–92, https://doi.org/10.1126/science.aaq0216.

8. Ian Johnston, "Women Must Have Equality with Men to Save the Planet, Experts Say," *Independent*, April 21, 2017, https://www.independent.co.uk/environment/women-men-equality-save-planet-say-experts-a7693781.html.

9. Francesca Gillett, "I Rebel So I Can Look My Grandchildren in the Eye," BBC News, October 15, 2019, https://www.bbc.co.uk/news/uk-50063449.

10. Ocean Conservancy, "Fighting for Trash Free Seas," https://oceanconservancy.org/trash-free-seas/, accessed February 7, 2020.

11. UN Environment Programme, "Champion of the Earth 2016: Afroz Shah," 2016, https://www.unenvironment.org/championsofearth/champion-earth-2016-afroz-shah.

12. NOAA Fisheries, "Catch and Release Best Practices," https://www.fisheries.noaa.gov/national/recreational-fishing/catch-and-release-best-practices, last updated July 22, 2019.

13. A species that is harmless in its native habitat may cause ecosystem disruption after it has been inadvertently transported to a new location. Take the example of the voracious European crab, now established along the coasts of North and South America, South Africa, Australia, and Japan, where it is devastating populations of indigenous shore life, including worms and shellfish. Invasive species are introduced in a number of ways, probably most commonly in a ship's ballast water, taken on board in one port and discharged in another.

14. International Seafood Sustainability Foundation, "Updated ISSF Report Rates State of Tuna Stocks Worldwide," 2019, https://iss-foundation.org/86-of-global-tuna-catch-continues-to-come-from-stocks-at-healthy-levels-but-some-stocks-remain-overfished/.

15. International Seafood Sustainability Foundation, "Updated ISSF Report."

16. Andrew Balmford, *Wild Hope: On the Front Lines of Conservation Success* (Chicago: University of Chicago Press, 2012).

17. World Wildlife Fund, "WWF Urges Marine Stewardship Council to Adopt Key Reforms," January 25, 2018, http://wwf.panda.org/wwf_news/press_releases/?321570/WWF-urges-Marine-Stewardship-Council-to-adopt-key-reforms.

18. Xanthe Clay, "No Cod, No Haddock—What Fish Can We Eat with a Clean Conscience?," *Daily Telegraph*, April 16, 2015, https://www.telegraph.co.uk/foodanddrink/11542239/No-cod-no-haddock-what-fish-can-we-eat-with-a-clean-conscience.html.

19. Jude Fuhnwi, "Women Are Championing Mangrove Conservation in Nigeria," Birdlife International, March 5, 2018, https://www.birdlife.org/worldwide/news/women-are-championing-mangrove-conservation-nigeria.

20. Rita Kant, "Textile Dyeing Industry an Environmental Hazard," *Natural Science* 4, no. 1 (2012): 22–26, http://dx.doi.org/10.4236/ns.2012.41004.

21. David Helvarg, foreword to *50 Ways to Save the Ocean* (Makowao, HI: Inner Ocean, 2006).

22. Emily Holden, "Nearly All Countries Agree to Stem Flow of Plastic Waste to Poor Nations," *Guardian*, May 11, 2019, https://www.theguardian.com/environment /2019/may/10/nearly-all-the-worlds-countries-sign-plastic-waste-deal-except-us.

23. Imogen Calderwood, "88% of People Who Saw 'Blue Planet II' Have Now Changed Their Lifestyle," Global Citizen, November 1, 2018, https://www.glo balcitizen.org/en/content/88-blue-planet-2-changed-david-attenborough/.

Chapter Ten

1. Michael Gross, "Q&A, Daniel Pauly," *Current Biology* 27 (2017): 399–407.

2. The Sea Around Us, "About," http://www.seaaroundus.org/, accessed February 7, 2020.

3. Daniel Pauly, *Vanishing Fish: Shifting Baselines and the Future of Global Fisheries* (Margate, UK: Greystone Books, 2019).

4. Daniel Pauly, *5 Easy Pieces: The Impact of Fisheries on Marine Ecosystems* (Washington, DC: Island Press, 2010).

5. Carl Walters, "Designing Fisheries Management Systems That Do Not Depend upon Accurate Stock Assessment," in *Reinventing Fisheries Management*, ed. Tony J. Pitcher, Paul J. B. Hart, and Daniel Pauly, Fish and Fisheries Series 23 (Dordrecht: Kluwer, 1998).

6. Pew, "Palau to Sign National Marine Sanctuary into Law," October 22, 2015, https://www.pewtrusts.org/en/about/news-room/press-releases-and-statemen ts/2015/10/22/palau-to-sign-national-marine-sanctuary-into-law.

7. Pew, "Palau Burns Illegal Fishing Boats from Vietnam," June 15, 2015, https:// www.pewtrusts.org/en/research-and-analysis/articles/2015/06/15/palau-ille gal-fishing.

8. Alan M. Friedlander et al., "Size, Age, and Habitat Determine Effectiveness of Palau's Marine Protected Areas," *PLoS ONE Biology* 12, no. 3 (March 30, 2017), https://doi.org/10.1371/journal.pone.0174787.

9. International Union for Conservation of Nature, "The Journey of Cook Islands: Marae Moana," June 26, 2018, https://www.iucn.org/news/oceania/201806/jour ney-cook-islands-marae-moana.

10. Crow White and Christopher Costello, "Close the High Seas to Fishing?," *PLoS ONE Biology* 12, no. 3 (March 25, 2014), https://doi.org/10.1371/journal.pbio .1001826.

11. U. Rashid Sumaila et al., "Winners and Losers in a World Where the High Seas Is Closed to Fishing," *Scientific Reports* 5 (2015), article 8481.

12. University of Oxford, University of York, and Greenpeace, "30X30 A Blueprint for Ocean Protection," April 2019, https://storage.googleapis.com/planet4-interna

tional-stateless/2019/03/479c73c5-30x30_blueprint_report_exec_summary
_web.pdf.

13. Catherine Benson Wahlen, "UK Government Calls for Protecting 30 Percent of
World's Oceans," International Institute for Sustainable Development, October 2, 2018, http://sdg.iisd.org/news/uk-government-calls-for-protecting-30
-percent-of-worlds-oceans/.

Index